Bridge to Abstract Mathematics

© 2012 by
The Mathematical Association of America (Incorporated)

Library of Congress Control Number: 2012943815

Print ISBN: 978-0-88385-779-3

Electronic ISBN: 978-1-61444-606-4

Printed in the United States of America

Current Printing (last digit):
10 9 8 7 6 5 4 3 2

Bridge to Abstract Mathematics

Ralph W. Oberste-Vorth
Indiana State University

Aristides Mouzakitis
Second Junior High School of Corfu

Bonita A. Lawrence
Marshall University

Published and distributed by
The Mathematical Association of America

Committee on Books
Frank Farris, *Chair*

MAA Textbooks Editorial Board
Zaven A. Karian, *Editor*

Richard E. Bedient
Thomas A. Garrity
Charles R. Hadlock
William J. Higgins
Susan F. Pustejovsky
Stanley E. Seltzer
Shahriar Shahriari
Kay B. Somers

MAA TEXTBOOKS

Bridge to Abstract Mathematics, Ralph W. Oberste-Vorth, Aristides Mouzakitis, and Bonita A. Lawrence
Calculus Deconstructed: A Second Course in First-Year Calculus, Zbigniew H. Nitecki
Combinatorics: A Guided Tour, David R. Mazur
Combinatorics: A Problem Oriented Approach, Daniel A. Marcus
Complex Numbers and Geometry, Liang-shin Hahn
A Course in Mathematical Modeling, Douglas Mooney and Randall Swift
Cryptological Mathematics, Robert Edward Lewand
Differential Geometry and its Applications, John Oprea
Elementary Cryptanalysis, Abraham Sinkov
Elementary Mathematical Models, Dan Kalman
An Episodic History of Mathematics: Mathematical Culture Through Problem Solving, Steven G. Krantz
Essentials of Mathematics, Margie Hale
Field Theory and its Classical Problems, Charles Hadlock
Fourier Series, Rajendra Bhatia
Game Theory and Strategy, Philip D. Straffin
Geometry Revisited, H. S. M. Coxeter and S. L. Greitzer
Graph Theory: A Problem Oriented Approach, Daniel Marcus
Knot Theory, Charles Livingston
Lie Groups: A Problem-Oriented Introduction via Matrix Groups, Harriet Pollatsek
Mathematical Connections: A Companion for Teachers and Others, Al Cuoco
Mathematical Interest Theory, Second Edition, Leslie Jane Federer Vaaler and James W. Daniel
Mathematical Modeling in the Environment, Charles Hadlock
Mathematics for Business Decisions Part 1: Probability and Simulation (electronic textbook), Richard B. Thompson and Christopher G. Lamoureux
Mathematics for Business Decisions Part 2: Calculus and Optimization (electronic textbook), Richard B. Thompson and Christopher G. Lamoureux
Mathematics for Secondary School Teachers, Elizabeth G. Bremigan, Ralph J. Bremigan, and John D. Lorch
The Mathematics of Choice, Ivan Niven
The Mathematics of Games and Gambling, Edward Packel
Math Through the Ages, William Berlinghoff and Fernando Gouvea
Noncommutative Rings, I. N. Herstein
Non-Euclidean Geometry, H. S. M. Coxeter
Number Theory Through Inquiry, David C. Marshall, Edward Odell, and Michael Starbird
A Primer of Real Functions, Ralph P. Boas
A Radical Approach to Lebesgue's Theory of Integration, David M. Bressoud
A Radical Approach to Real Analysis, 2nd edition, David M. Bressoud
Real Infinite Series, Daniel D. Bonar and Michael Khoury, Jr.
Topology Now!, Robert Messer and Philip Straffin
Understanding our Quantitative World, Janet Andersen and Todd Swanson

MAA Service Center
P.O. Box 91112
Washington, DC 20090-1112
1-800-331-1MAA   FAX: 1-301-206-9789

Contents

Some Notes on Notation . . . xi

To the Students . . . xiii
    To Those Beginning the Journey into Proof Writing xiii
    How to Use This Text xiv
    Do the Exercises! xiv
    Acknowledgments xv

For the Professors . . . xvii
    To Those Leading the Development of Proof Writing
        for Students in a Broad Range of Disciplines xvii

I THE AXIOMATIC METHOD   1

1 Introduction   3
    1.1 The History of Numbers 3
    1.2 The Algebra of Numbers 4
    1.3 The Axiomatic Method 6
    1.4 Parallel Mathematical Universes 8

2 Statements in Mathematics   9
    2.1 Mathematical Statements 9
    2.2 Mathematical Connectives 11
    2.3 Symbolic Logic 16
    2.4 Compound Statements in English 20
    2.5 Predicates and Quantifiers 21
    2.6 Supplemental Exercises 26

3 Proofs in Mathematics   29
    3.1 What is Mathematics? 29
    3.2 Direct Proof 30
    3.3 Contraposition and Proof by Contradiction 33
    3.4 Proof by Induction 37
    3.5 Proof by Complete Induction 42
    3.6 Examples and Counterexamples 46
    3.7 Supplemental Exercises 48
    How to THINK about mathematics: A Summary 52

How to COMMUNICATE mathematics: A Summary 52
How to DO mathematics: A Summary 52

II SET THEORY 53

4 Basic Set Operations 55
4.1 Introduction 55
4.2 Subsets 56
4.3 Intersections and Unions 58
4.4 Intersections and Unions of Arbitrary Collections 61
4.5 Differences and Complements 64
4.6 Power Sets 65
4.7 Russell's Paradox 66
4.8 Supplemental Exercises 68

5 Functions 75
5.1 Functions as Rules 75
5.2 Cartesian Products, Relations, and Functions 76
5.3 Injective, Surjective, and Bijective Functions 82
5.4 Compositions of Functions 83
5.5 Inverse Functions and Inverse Images of Functions 85
5.6 Another Approach to Compositions 87
5.7 Supplemental Exercises 89

6 Relations on a Set 93
6.1 Properties of Relations 93
6.2 Order Relations 94
6.3 Equivalence Relations 98
6.4 Supplemental Exercises 103

7 Cardinality 107
7.1 Cardinality of Sets: Introduction 107
7.2 Finite Sets 108
7.3 Infinite Sets 110
7.4 Countable Sets 113
7.5 Uncountable Sets 114
7.6 Supplemental Exercises 118

III NUMBER SYSTEMS 121

8 Algebra of Number Systems 123
8.1 Introduction: A Road Map 123
8.2 Primary Properties of Number Systems 123
8.3 Secondary Properties 126
8.4 Isomorphisms and Embeddings 128
8.5 Archimedean Ordered Fields 129

    8.6 Supplemental Exercises 132

9 The Natural Numbers   137
    9.1 Introduction 137
    9.2 Zero, the Natural Numbers, and Addition 138
    9.3 Multiplication 141
    9.4 Supplemental Exercises 143

Summary of the Properties of the Nonnegative Integers   144

10 The Integers   145
    10.1 Introduction: Integers as Equivalence Classes 145
    10.2 A Total Ordering of the Integers 146
    10.3 Addition of Integers 147
    10.4 Multiplication of Integers 149
    10.5 Embedding the Natural Numbers in the Integers 151
    10.6 Supplemental Exercises 152

Summary of the Properties of the Integers   153

11 The Rational Numbers   155
    11.1 Introduction: Rationals as Equivalence Classes 155
    11.2 A Total Ordering of the Rationals 156
    11.3 Addition of Rationals 157
    11.4 Multiplication of Rationals 158
    11.5 An Ordered Field Containing the Integers 159
    11.6 Supplemental Exercises 161

Summary of the Properties of the Rationals   164

12 The Real Numbers   165
    12.1 Dedekind Cuts 165
    12.2 Order and Addition of Real Numbers 166
    12.3 Multiplication of Real Numbers 168
    12.4 Embedding the Rationals in the Reals 169
    12.5 Uniqueness of the Set of Real Numbers 169
    12.6 Supplemental Exercises 172

13 Cantor's Reals   173
    13.1 Convergence of Sequences of Rational Numbers 173
    13.2 Cauchy Sequences of Rational Numbers 175
    13.3 Cantor's Set of Real Numbers 177
    13.4 The Isomorphism from Cantor's to Dedekind's Reals 178
    13.5 Supplemental Exercises 180

14 The Complex Numbers   181
    14.1 Introduction 181
    14.2 Algebra of Complex Numbers 181
    14.3 Order on the Complex Field 183
    14.4 Embedding the Reals in the Complex Numbers 184

14.5 Supplemental Exercises 185

IV TIME SCALES 187

15 Time Scales 189
15.1 Introduction 189
15.2 Preliminary Results 189
15.3 The Time Scale and its Jump Operators 191
15.4 Limits and Continuity 193
15.5 Supplemental Exercises 197

16 The Delta Derivative 199
16.1 Delta Differentiation 199
16.2 Higher Order Delta Differentiation 203
16.3 Properties of the Delta Derivative 204
16.4 Supplemental Exercises 208

V HINTS 209

17 Hints for (and Comments on) the Exercises 211
Hints for Chapter 2 211
Hints for Chapter 3 211
Hints for Chapter 4 212
Hints for Chapter 5 213
Hints for Chapter 6 214
Hints for Chapter 7 214
Hints for Chapter 8 215
Hints for Chapter 9 216
Hints for Chapter 10 216
Hints for Chapter 11 217
Hints for Chapter 12 218
Hints for Chapter 13 219
Hints for Chapter 14 220
Hints for Chapter 15 220
Hints for Chapter 16 220

Bibliography 223

Index 225

About the Authors 231

Some Notes on Notation

We use this notation for commonly used sets of numbers:

$\mathbb{N}$ is the set of natural numbers $\{1, 2, 3, 4, 5, \ldots\}$
$\mathbb{Z}_n$ is the set of the first n integers $\{1, 2, 3, \ldots, n\}$
$\mathbb{Z}$ is the set of integers
$\mathbb{Z}_+$ is the set of nonnegative integers $\{0, 1, 2, 3, 4, \ldots\}$
$\mathbb{Z}^*$ is the set of nonzero integers
$\mathbb{Q}$ is the set of rational numbers
$\mathbb{Q}_+$ is the set of nonnegative rational numbers
$\mathbb{Q}_+^*$ is the set of positive rational numbers
$\mathbb{R}$ is the set of real numbers
$\mathbb{R}_+$ is the set of nonnegative real numbers
$\mathbb{R}_+^*$ is the set of positive real numbers
$\mathbb{C}$ is the set of complex numbers
(a, b) is the open interval $\{x \in \mathbb{R} \mid a < x < b\}$
$[a, b]$ is the closed interval $\{x \in \mathbb{R} \mid a \leq x \leq b\}$
$(a, b]$ is the interval $\{x \in \mathbb{R} \mid a < x \leq b\}$
$[a, b)$ is the interval $\{x \in \mathbb{R} \mid a \leq x < b\}$

Most authors use $\mathbb{Z}_n = \{0, 1, 2, \ldots, n-1\}$. The choice here is more natural in our discussion of cardinality in Chapter 7.

We use this notation for subsets and proper subsets:

$A \subset B$ A is a subset of B (A may equal B)
$A \subsetneq B$ A is a proper subset of B ($A \neq B$)

We never use $A \subseteq B$. Many authors use $A \subseteq B$ for subsets and $A \subset B$ for proper subsets. When you read "$A \subset B$" somewhere, be careful to determine its meaning.

We use this notation for functions:

$f : X \to Y$ f is a function from domain X to codomain Y
$f|_S$ the restriction of f to a subset S of its domain

xi

The symbol □ will denote the end of a proof; it may be read as "Q.E.D." abbreviating the Latin

quod erat demonstrandum,

which translates the Greek of Euclid of Alexandria

ὅπερ ἔδει δεῖξαι.

To the Students

'Mathematicians,' [Uncle Petros] continued, 'find the same enjoyment in their studies that chess players find in chess. In fact, the psychological make-up of the true mathematician is closer to that of the poet or the musical composer, in other words of someone concerned with the creation of Beauty and the search for Harmony and Perfection. He is the polar opposite of the practical man, the engineer, the politician or the ...' — he paused for a moment seeking something even more abhorred in his scale of values — '...indeed, the businessman.' — Apostolos Doxiadis [3]

To Those Beginning the Journey into Proof Writing

As undergraduates in the 1970s, we, the authors of this text, learned how to construct proofs essentially by osmosis. We learned much of the material contained in this text in much the same way. That is, we learned the art of proof mostly by imitating the experts. We listened to and watched their presentations. We read their textbooks. For us, the process of learning how to prove theorems began in the calculus sequence and continued through linear algebra and beyond.

This method was similar to learning to sketch or paint a beautiful landscape by imitating the techniques and vision of a master. While such modeling has its place in the development of learning to construct a proof, it is merely one of the first steps toward the creation of your own masterpiece. This text is designed to offer you the tools you will need to create proofs that are both logical and valid and, equally important, to encourage you to create from your own perspective. We hope to bridge the gap between observation and imitation of the work of one who is well-versed in proof writing, and striking out on your own, tools in hand, on your own path that takes you to your masterpiece.

You may have already had some experience observing presentations of well-constructed proofs in a geometry or calculus class. Perhaps you were asked to recreate these proofs as part or your examinations. For those majoring in mathematics, awaiting you in upper-level courses, such as modern and linear algebra, and real and complex analysis, is the expectation of your ability to prove fundamental results and use them to prove further results. Regardless of your field of study, proofs offer valuable insight into theoretical structure and therefore the ability to construct a proof gives the proof-writer the power to see below the surface and observe the underpinnings of an idea. We often wonder how our own mathematical development would have been altered had we taken a bridge course before we dove into such upper level courses in our mathematical training. We hope that you—today's students of logic and proof-writing from a broad spectrum of disciplines—find this type of textbook useful.

The book has a dual purpose. The primary goal is to guide you in the process of writing proofs. We use, as the medium for your proof constructions, a collection of what we hope are interesting pieces of mathematical content: our secondary goal is for you to get actively involved in order to assimilate this content. As important as having intriguing topics to work with as you develop your proof-writing skills is the process of transforming your thought processes from passive to active and from computational to creative. This is a fundamental step in the development of a foundation from which you can learn any body of knowledge that has an axiomatic structure.

How to Use This Text

The text has five parts. Part I contains introductory material presented in three chapters. Chapter 1 provides some historical background concerning the invention and development of mathematics. In Chapter 2, we define a mathematical statement and use basic symbolic logic to systematize the principle of valid reasoning. Questions we will be concerned with are: How can mathematical statements be combined? When is truth preserved when statements are combined in specific ways characteristic of mathematical thinking? When are two statements equivalent? Chapter 3 presents approaches and techniques commonly used to prove mathematical statements. The techniques will serve as a bridge from low-level cognitive processes, characteristic of computational skills and procedural understanding, to higher-level cognitive processes, focused on conceptual understanding and creative constructions.

The medium you will use to hone your proof-writing skills, the mathematical content, begins in Part II. Chapter 3 deals with some basic proof methods and techniques. This theoretical machinery is used in the study of set theory in Chapter 4. Then it is extended to study functions from one set to another in Chapter 5, relations on a set in Chapter 6, and cardinality of a set in Chapter 7.

Part II—particularly Chapters 4, 5, and 6—builds a foundation for writing many kinds of proofs. Parts III and IV are more difficult, at least in some places. They provide preparation for abstract algebra and advanced calculus (or analysis) courses.

Part III leads you through the construction and examination of various number systems. Chapter 8 offers important properties and relevant proofs related to the development of number systems. Chapters 9 through 14 deal with the sets of natural (or counting or whole) numbers, integers, rational numbers (or fractions), real numbers, and complex numbers.

In physics, you may have functions that describe the position, velocity, and acceleration of a moving object. These quantities may be considered instantaneously in time or on average in a time period. In Part IV we present a structure that allows us to study the properties of such functions and their generalizations. If you have studied calculus, you may know something about this, but we will be more general and precise.

Part V provides hints for the exercises found in the chapters. This is valuable resource for you. Use it wisely. Ultimately our goal is to give you the tools, opportunity, and encouragement to create proofs of your own.

Do the Exercises!

Within each chapter, definitions are presented with examples and comments. Perhaps it is possible to be able to reproduce some mathematical proofs merely by watching, listening, and reading. However, we believe that you will be better off learning actively, that is, by supplying many of the proofs yourself. The text takes small steps forward; proofs are provided at larger leaps.

As you progress through the book, begin each exercise by trying to find a logical path to the conclusion. Only after you have spent time trying to find a plan for your proof should you look at the hint. If you begin by looking at the hint, you will deprive yourself of the joy and excitement that comes with finding your own way; you may discover a different and more interesting path than the one we suggest in the hint. If all else fails, you can always ask the experts.

Do not overlook the resource of your fellow students. Listening to their ideas about how to proceed with a proof and seeing what they take from your ideas is invaluable. It is difficult to do mathematics without discussing it with someone, even a non-expert. It is proper to challenge the ideas your fellow students tell you about their proofs. Free discourse among students and colleagues is the key to developing your own creative style. Do not be satisfied with finding a single path to a result and do not back off your idea until you find that there is a missing link in your argument.

Finally, enjoy yourself creating your own masterpieces!

Acknowledgments

We would like to thank the Fall 2000, Spring 2001, and Summer 2001 classes of MGF 3301 taught by the first author at the University of South Florida for testing drafts of this text. Their corrections and suggestions have been invaluable. In particular, we would like to thank Paul Anderson, Ray Burrus, Christie Burton, Leon Calleja, Thuc Cao, Nathan Chau, Teresa Chung, Jason Copenhaver, Mindy Eason, Adam Francis, Alynne Frewin, Russell Gerbers, Bridget Giroux, Erika Johnson, Kristy Kazemfar, Sarah Lahlou-Amine, Christopher Ledwith, Carson McCoy, Rose Nestor, Cheryl Ng, Ryan Parrish, Michelle Richardson, Patrick Robbins, Emily Roberts, Cheryl Scilex, Anthony Upchurch, Adrianne Waltz, Jeanne Waser, and Aimee Yates. We would also like to thank professors Edwin Clark and Boris Shekhtman for using drafts of the text at the University of South Florida from 2001 to 2003.

We also thank students in MTH 300 at Marshall University who have used versions of this text since 2003. We thank our former students, including Ann Capper, Laura Caskey, Courtney Green, Victor Imperi, Matthew Lucas, Shannon Miller, Michael Pemberton, Bonnie Shook, Erin Simmons, John Stonestreet, Justin Wince, Shawn Cotton, Derek Musgrave, Tu Nguyen, Lance Perry, Samantha Skelley, James Wroten, Adam Chain, James Cox, Stephen Deterding, Lauren Keller, Mallory Price, Tyler Torlone, Brittany Whited, Kayla Chappelle, Cecylia Dembinski, Kelsi Halbert, Lindsay Hansen, Rebecca Hovemeyer, David Poole, David Sargent, Amanda Sellers, and Patrick Stewart for helpful comments and corrections. We especially thank professors Basant Karna and Carl Mummert for using versions of this text in 2009 and 2010.

Further thanks go to the students in MATH 380 at Indiana State University during Spring 2012, including Nancy DeGott, Chandra Hull, Sidney Stines, and Elijah Waterman.

For the Professors

To Those Leading the Development of Proof Writing for Students in a Broad Range of Disciplines

The authors of this book were trained in the process of proof writing using the 1970s method of listen, observe, read, and imitate. We were introduced to the art of proof early in our calculus sequences and we developed our skills by studying the techniques used by our professors and presented in our textbooks.

Undergraduate mathematics curricula in the U.S. have undergone some changes since then. The most visible change may be the end product of Calculus Reform. With its integration of numerical, analytical, and geometrical aspects of the subject and the integration of computing technology, some topics have been removed or minimized in the curricula. We saw many proofs in our calculus courses, even if we were not always held responsible for them on the examinations. This is not always the case today. In fact, at many institutions a variety of calculus sequences exists and textbooks used in some do not give any proofs.

Calculus Reform is not the only force driving changes in the curricula. The growing number of high school graduates who continue their educations at post-secondary institutions has put pressure on the calculus curricula by requiring increasingly extensive reviews of algebra and trigonometry. Also, changes in major often leads a future major in mathematics to take a calculus course designed for other majors.

As a result of these fundamental changes in calculus courses, students have been enrolling in course such as linear algebra with weaker backgrounds in construction and understanding proofs. Recently, many institutions have inserted courses into their curricula to address this problem. These courses go by many names. The choice of mathematical content is not universal, but the prime objective is universal. That objective is to prepare students to deal with proof in their later courses. For example, at the University of South Florida, the faculty named their course *Bridge to Abstract Mathematics*. Amazingly, the title was more contentious that the course itself!

The usual content for "Bridge" courses is set theory and related topics. Courses in set theory were common part of the mathematics major earlier in the twentieth century. Between the 1950s and the 1970s most of them were squeezed out of the requirements of the mathematics major by courses in linear algebra and abstract algebra, and a relaxation of requirements in favor of elective courses.

The text has two purposes: to present mathematical content and to guide students on how to create proofs. We hope the content is interesting to your students, but the process of transforming the thought processes from a passive computational orientation to an active creative orientation is more important.

The text has five parts. Part I contains introductory material presented in three chapters. Chapter 1 provides some historical background concerning the invention and development of different kinds of numbers, and gives a brief introduction of the axiomatic method and its scientific and pedagogical merits.

Chapter 2 introduces symbolic logic, which takes a formal approach in building theory with a goal to systemize and codify the principles of valid reasoning. The term "formal" means that we are not concerned with the content and meaning of statements, but exclusively with their form.

Chapter 3 illustrates specific techniques and methods of mathematical proof that are common to mathematical practice and deserve an explicit treatment.

The mathematical content begins in Part II, consisting of Chapters 4 through 7. Chapter 4 presents the elements of set theory, which has been crucial for the modern development of mathematics. It fulfills three important functions. (See [10].) First, set theoretic language permeates a large part of mathematics, supplying its diverse areas with common modes of reasoning. Second, it plays the foundational role of supplying the subject matter of mathematics. Third, it is an important tool used in the study of the infinite.

Chapter 5 is concerned with the fundamental concept of function. Here we formal prove and ask for proofs of some basic facts about functions that are necessary in many areas of mathematics.

Chapter 6 formalizes the concept that two mathematical objects are related according to a specific common characteristic. We explain how a relation on a set gives rise to the generalized notion of order. The important concept of an equivalence relation is given elaborate treatment. This notion is used in the construction of number systems in Part III.

In Chapter 7 we give an elementary treatment of cardinality that extends the notion of number to infinite sets. The relevant theory was primarily produced by the genius of Georg Cantor and contains some landmark results, some of which are included here.

Beginning with Chapter 8 in Part III, where the terminological equipment and some relevant proofs concerning the primary and secondary properties of number systems are laid out, we are is concerned with the construction of number systems. We construct the natural numbers, the integers, the rational numbers, the real numbers, Cantor's real numbers, and the complex numbers in Chapters 9 through 14.

It is our opinion that an understanding of such systems at a fundamental level is a necessary prerequisite for a deeper grasp of mathematical analysis. Usually the real number system with its associated properties is taken for granted in undergraduate courses. Here we give two different constructions of it that present real numbers differing in ontological flavor until we prove they are identical up to isomorphism. Pedagogically, the inclusion of the construction of number systems offers an opportunity for practicing proof and mathematical rigor. The theoretical developments in these chapters culminate with a proof of the existence and uniqueness of a complete ordered field.

Part IV is an introduction to time scale calculus, a structure that unifies discrete and continuous analysis. The benefits of a study of this new area of mathematics at this level is the students' sense of familiarity with the concepts discussed that stems from their study of calculus on the real intervals. Some familiar properties of functions are generalized to functions defined on any closed subset of the real line.

Leading the Development of Proof Writing

Part V provides some hints to the exercises contained with the main narrative of each chapter. The goal is to provide a resource for students to use only after having spent sufficient time trying to formulate a plan of their own. We do not want to deprive students of the chance to offer a different approach than the one we had in mind and we encourage them to create more than one proof for a given statement.

The goal of this text is to prepare students for the standard mathematics curriculum. The content of the text has little overlap with the standard curriculum. Parts of Part II are generally expected as common knowledge. Part III gives students preparation for an abstract algebra course. Part IV prepares students for a rigorous treatment of calculus or vector analysis.

Within each chapter, definitions are presented with examples and comments. The proofs of propositions and theorems are often left as exercises. To a certain extent one can learn how to prove theorems merely by watching, listening, and reading proofs produced by others. However, we are convinced that active personal involvement and initiative with writing proofs has a profound impact on the development of students ability to create their own arguments. Such development is encouraged by the format of the text, taking small steps forward, with some proofs provided where they involve a larger leap forward. We encourage students to discuss their ideas with their peers and to question ideas presented by everyone involved in these mathematical discussions.

The chapters on the constructions of the natural numbers and the real numbers, Chapters 9 and 12, respectively, are considerably more difficult that the other chapters of Part III. They can be skipped as long as their properties are understood for use in the other chapters. Enjoy and let us know what you think (oberstevorth@gmail.com).

I

THE AXIOMATIC METHOD

Introduction

1.1 The History of Numbers

The seemingly easy concept of number is an abstraction that came late in our intellectual evolution. The ancient shepherd had no method of counting his sheep. He used the more primitive concept of a one-to-one correspondence instead, perhaps by notching a bone as many times as the number of his sheep.

The counting numbers or natural numbers

$$1, 2, 3, 4, \ldots$$

were the first to appear . Pythagoras of Samos, a mystic who lived in the sixth century B.C. and who is famous for the theorem that bears his name, considered the natural numbers to be the stuff out of which the universe is made.

The positive rational numbers, which come up when we divide a whole thing into smaller pieces, were invented next. They constitute a natural extension of the natural numbers. A number is positive rational if and only if it can be expressed as a quotient of natural numbers. Such quotients are also called fractions.

Pythagoras and his disciples knew about fractions. However, the discovery of numbers that cannot be expressed as fractions shook the philosophical foundations of the Pythagorean school and, at the same time, caused the first big crisis in mathematics. The dramatic discovery of what came to be called *irrational* numbers is reflected in the story of Hippasus according to which this notable Pythagorean philosopher was drowned at sea when he disclosed the incommensurability of the side of a square with its diagonal. This unhappy event inspired a modern novel [8].

In response to this crisis, Eudoxus of Cnidus devised an ingenious theory of irrational quantities that anticipated the rigorous treatment given by Georg Cantor and Richard Dedekind in their constructions of the real number system almost two and one half millennia later. This is perhaps the first historical example in the development of mathematics where a critical moment provided fertile ground for the enrichment of mathematics and opened new horizons.

The number zero was introduced by the Hindus in the ninth century A.D., but some historians conjecture that it was known earlier by the Babylonians. The negative and complex numbers are related because of the definition of imaginary numbers as the square roots of negative numbers. Negative and complex numbers were slowly accepted. However, the existence of negative and complex numbers was still often viewed as absurd for a millennium, despite their increased formal use over the centuries, particularly in algebra.

Economic developments—for example, an international banking system spreading out of Italy in the fourteenth century as Europe was moving from the Middle Ages to the Renaissance—provided favorable circumstances for the wider acceptance of negative numbers. Their use was perhaps influenced by Leonardo Fibonacci in the thirteenth century; he noted that a negative number could be regarded as a loss whereas a positive number could represent a gain.

The universal acceptance of the complex numbers and, hence, the negative numbers, did not occur until the 1830's. The early part of the nineteenth century saw developments in algebra, particularly the extensions of the naturals to the integers and rationals and of the reals to the complex numbers. The notion of equivalence classes stood at the heart of these developments, though this language was not developed until the twentieth century. In 1830, George Peacock published what is now considered one of the first books on abstract algebra [9]. Several years later, he was careful to separate "arithmetical" from "symbolical" algebra where the former does not allow negative numbers.

The real numbers, at least the positive real numbers, were accepted since the time that Eudoxus developed the irrational numbers. However, the axiomatic construction of the real numbers did not occur until the nineteenth century. Many mathematicians had attempted to put the reals on a solid foundation. It was finally achieved by Dedekind and Cantor in the 1870's.

This brings us back to the naturals. The most natural of the numbers were the most difficult, requiring the axioms of set theory that were taking form at the beginning of the twentieth century. Today, we give Richard Dedekind, Giuseppe Peano, Bertrand Russell, and Alfred North Whitehead the credit for formalizing what was known and used for ages (see [11]).

We encourage you to explore the history of mathematics through coursework or reading. Among the many texts are [1], [6], and [7].

1.2 The Algebra of Numbers

Apart from any social conditions that may have played a role in the development of number systems, there also exist intrinsic reasons for their development that are peculiar to the inner dynamics of mathematical activity. The step from natural numbers to rational numbers derives from human activity where an apple and yet another make two apples and where two people sharing an apple make for one half of an apple for each. Irrational numbers are deeper; in the time of Eudoxus, rational numbers were encountered in geometrical issues such as the lengths of diagonals of a rectangle whose sides had lengths expressed by natural numbers. Negative numbers are understandable, as by Fibonacci above, if not as easily realizable. Where do complex numbers come from?

Pause for a moment to reflect on the names given to the expanding sets of numbers. They surely reflect the prejudices of the namers or, at the very least, the harsh criticism their introduction produced: positive versus negative, rational versus irrational, and real versus imaginary.

There is an algebraic way of explaining the need for most of these numbers and it

1.2. The Algebra of Numbers

is this algebraic need that first pointed to the complex numbers. We will give successive elementary equations in terms of successive sets of known numbers whose solutions are not in those sets, but in successively expanded sets.

Suppose that
$$m + x = n$$
for natural numbers m and n. Then exactly one of the following holds:
$$n > m \quad \text{or} \quad n = m \quad \text{or} \quad n < m.$$

If $n > m$, then the equation has a solution in the natural numbers, which is called $n - m$. If $n = m$, then the existence of 0 is suggested. Finally, if $n < m$, then the negative integers are needed for the equation to have a solution. Equations of the form
$$qn = m,$$
for integers m and n, with $n \neq 0$, lead us to the rational numbers. Finally, there exist polynomial equations of the form
$$a_n x^n + a_{n-1} x^{n-1} + a_{n-2} x^{n-2} + \cdots + a_1 x + a_0 = 0,$$
with coefficients that are integers, that lead to the irrational numbers and the complex numbers. For example,
$$x^2 - 2 = 0$$
leads us to the irrational numbers and
$$x^2 + 1 = 0$$
leads to the imaginary numbers.

The set that is missing from the above discussion is the set of real numbers whose construction can be viewed as a completion of the rational numbers. Prior to the full explanation and rigorous treatment of the real number system that was realized in the nineteenth century, numbers that are real solutions of polynomial equations with integer coefficients were given special consideration; these are called algebraic numbers. Also, proofs that some real numbers, such as π, are not algebraic were supplied before the real numbers were rigorously defined. These numbers are called transcendental.

Rational numbers can be written as repeating decimals such as
$$\frac{1}{1} = 1.\overline{0}\ldots,$$
$$\frac{1}{2} = 0.5\overline{0}\ldots,$$
and
$$\frac{1}{3} = 0.\overline{3}\ldots.$$
Remarkably, some rational numbers have two representations such as
$$1.\overline{0}\cdots = 0.\overline{9}\ldots$$

and
$$0.5\overline{0}\cdots = 0.4\overline{9}\ldots.$$
To justify the former, let
$$x = 0.99\overline{9}\ldots.$$
Multiply both sides of the equation by 10 and subtract the first from the second to get $9x = 9$. You might have some reservations as to how the two right hand sides are subtracted. Perhaps this is your first difficulty with encountering infinity.

We can imagine other decimals where the digits need not form any pattern. We call the set of all decimals the real numbers. Making this precise is not as easy as it may appear!

1.3 The Axiomatic Method

Mathematics is highly respected in our culture. It inspires awe or fear in most people. To a few it is a source of joy and admiration, an artistic game of the highest quality, the language in which the book of nature is written.

Whence mathematics derives its power? Is there a royal road to it? Is it accessible to all mortals? Mastering it requires hard work, discipline, concentration, self-respect, and self-confidence. Armed with these qualities, you have to plunge into this book and be actively engaged in deciphering its content. That means that you have to ask yourself questions and, by trying to answer them, ask more questions; you have to try to discover the path that the intellect has followed to derive a proof, a path that is now hidden either for aesthetic purposes or for the sake of precision and simplicity.

Mathematics started as an empirical discipline in Mesopotamia and Egypt to serve everyday practical needs. Rules were invented to measure distances, areas, and volumes, thus constituting an embryonic form of the body of knowledge called geometry.

With the Greeks geometry acquired qualitatively different character. With Thales of Miletus as a pioneer and then with the school of Pythagoras, an effort was initiated to establish geometry as an an abstract theoretical discipline. The culmination of this process was the work of Euclid of Alexandria in which the deductive method was refined and reached a high level of perfection even judged by contemporary standards. The axiomatic approach encapsulates those qualities for which mathematics is mostly praised, namely the power of abstraction, linguistic precision, lack of ambiguity, systematization of thought, and transparency of argumentation. Since the axiomatic method still prevails as a way to justify the results of the mathematicians' creative imagination, let us describe and explain its main features.

In ordinary discourse, words are rarely defined. Instead, the meaning of words is usually given implicitly by the socio-cultural and situational context within which they are uttered. This often results in ambiguity, particularly when abstract concepts are involved, insofar as each person acquires an understanding of a word from the totality of instances the word has been encountered. Dictionary definitions are often cyclical.

In the world of mathematics—of idealized abstract entities—not all words can be defined. To determine the meaning of a word, this word has to be explained in terms of other words and this process will never terminate unless we reach a word whose meaning is generally accepted without explanation. A word whose meaning is generally accepted is called

1.3. The Axiomatic Method

a *primitive term* or an *undefined term*. The terms *point*, *straight line*, and *plane* are examples of primitive terms used in geometry. Other terms of the system in question are defined in terms of previously defined terms and the primitive terms.

When starting from scratch, i.e., from the level of undefined terms, it is not easy to give definitions of even simple objects. However, this is necessary since we cannot assume that different people understand terms the same way. Disagreements exist even in the mathematical literature. Take the basic geometric notion of a triangle. See Figure 1.1.

Figure 1.1. Triangle vs. triangular region

You may think everyone knows what a triangle is. Try this experiment: ask a few non-mathematicians "What is a triangle?" You will probably get some different answers, perhaps including drawings. The most popular answer may describe a triangular region such as the shaded region in Figure 1.1. A geometer may call that a region bounded by a triangle and give the unshaded figure as an example of a triangle! Hence the need for definitions, even for simple mathematical objects.

If, as we explained above, the primitive terms are not defined in an axiomatic system, how are their meanings determined? Are we to construe them as we please? From the axiomatic point of view the undefined terms are implicitly defined by basic propositions that involve these terms. Such propositions are called *axioms*. Diverging from Euclid's traditional approach, which lasted for two millennia and according to which the axioms were accepted as self-evident truths, the modern axiomatic approach considers the axioms as unproved propositions that are necessary for the proof of all other propositions. But why are axioms needed? Suppose you state an opinion on an issue using a single sentence, for example "freedom is preferable to happiness" and your interlocutor asks you to justify it. After your next sentence he asks you to justify this sentence also. It is clear that this process will never end, unless you and your interlocutor agree upon the truth of some sentence without justification.

On the basis of axioms and rules of logic, mathematicians masterfully erect the edifice consisting of a series of proved propositions. The historical prototype for the axiomatic approach is Euclids *Elements*, written around 300 B.C. From a pedagogical point of view, by presenting a multitude of isolated mathematical facts into a unified coherent whole, the axiomatic approach provides a framework for organizing and ordering learning activities enriching and enhancing their meaning.

In mathematics, *proof* consists of a sequence of logical steps by which we deduce the truth of a statement from axioms that are explicitly stated or from propositions previously derived. Proof is the backbone of mathematics and apart from justifying the truth of mathematical statements it also fulfills a variety of educational functions. Drawing on Michael De Villiers, [2], we elaborate on the most important pedagogical functions of proof.

Proof might illuminate a mathematical result by giving an insight into why it is true. For example, deriving the quadratic formula throws light on why it works, whereas taking it for granted leaves it mysterious. That encourages blind submission to authority. This is the case if the omission of proof is not a curricular exception but the rule.

Proof is also a means of communication, providing a forum for productive dialogue and critical debate. If the proof process is taking place in the classroom, the power of an argument is under public scrutiny and is in harmony with the actual mathematical practice among the members of the mathematical community. This is true even when the dialogue is imaginary between the reader and the author of a book.

Proof constitutes an intellectual challenge that may serve the function of self-realization and self-fulfillment as a result of diligence, hard work, and perseverance. Since antiquity, throughout the centuries mathematicians have struggled with problems that at times posed them insurmountable difficulties. Although they were not always successful in their efforts, their preoccupations with even insoluble problems were particularly productive in terms of discovering and inventing new mathematical results and opening new horizons for mathematics.

Mathematicians have contrived various methods of deducing a desired result, that is, they have devised different kinds of proof. We illustrate the methods with specific examples in the next chapter. However, there is more to the methodology of doing proofs than the types of proofs. This involves the thought processes themselves.

This book is intended to display a good part of the elementary mathematical machinery for tackling and solving mathematical problems. Students learn to acquire, develop, and refine this equipment through practice.

1.4 Parallel Mathematical Universes

One question that arises when an area of mathematics is to be founded on an axiomatic basis is whether there is a unique set of axioms that can serve. It turns out that two different collections of axioms can lead to the same mathematics, that is, to the same collection of true statements. Here truth means derivability of a statement from the original axioms. It is also the case that the removal, addition, or other change of just one axiom can dramatically change the mathematics.

You may already know the story of Euclid's Parallel Postulate from your high school geometry class. (We regard the words axiom and postulate as synonyms, though Euclid and others regarded them as different in a technical sense.) In plane geometry, the Parallel Postulate is equivalent to the statement that there is a unique line that is parallel to a given line passing through a point not on the original line. One may instead assume that no parallel line may exist or that more than one parallel line may exist. Such changes in the axioms lead to new and different geometries. From a mathematical point of view, these non-Euclidean geometries are as legitimate as Euclidean geometry.

The issue of which geometry to use to model physical space has to be decided empirically. Newtonian physics tried to model the physical universe with Euclidean geometry. After the work of Einstein on relativity, the accepted model of space (or more precisely space-time) is based on non-Euclidean geometry.

Statements in Mathematics

2.1 Mathematical Statements

Whatever mathematics may be as a mental activity, it is communicated as a language. Therefore, it has its specific syntax, its own technical terms, and its own conventions. Mathematics is also an exact science, which means that we are obliged to express our mathematical thoughts with high precision. A deviation from the norm may lead to a complete distortion of the intended meaning.

The purpose of this chapter is to make explicit a number of principles that pertain to mathematical logic. This branch of mathematics formalizes the principles of mathematical reasoning, principles that permeate mathematical thinking, be it consciously or subconsciously.

In mathematics, we assert the truth of some statements. Other statements are to be proved or disproved. This is a fundamental dichotomy:

> No mathematical statement is both true and false.

Viewed axiomatically, you can take *true* and *false* to be undefined terms; the above dichotomy can be taken as an axiom. We should clarify what is meant by a mathematical statement.

Definition 1. A declarative sentence is a *logical statement* if and only if it is unambiguously either true or false. A *mathematical statement* is a logical statement used in mathematical discourse.

By *statement*, without an adjective, we will always mean a logical or, usually, a mathematical statement. Here are some examples.

The sentence

> The number 2 is positive.

is a true mathematical statement and the sentence

> The number 2000 is divisible by 3.

is a false mathematical statement. Note that the previous two sentences are both declarative.

By definition, sentences that are not declarative cannot be logical statements; such sentences are neither true nor false. An interrogative sentence such as

> Who is there?

is not a logical statement. An imperative sentence such as

$$\text{Go away.}$$

is not a logical statement. An exclamatory sentence such as

$$\text{Wow!}$$

is not a logical statement. A sentence like

$$\text{Hello.}$$

is not a logical statement either.

A declarative sentence may be a logical statement even if its truth value is not known. This is common in the most advanced areas of mathematics; it is the fuel of mathematical research. The following sentence

$$\text{The } 10^{10}\text{th digit in the decimal expansion of the number } \pi \text{ is 3.}$$

is a mathematical statement. Even if no one knows whether this is true or false, it must be exactly one or the other. This is what we mean by unambiguous in Definition 1.

Not every declarative sentence is a statement. Opinions are not logical statements. For example, the declarative sentence

$$\text{Vanilla is the best ice cream flavor.}$$

is not a logical statement since it does not have a truth value; it is true that a person may have an opinion on this subject, but this does not make the statement true or false. On the other hand,

$$\text{Vanilla is Jon's favorite ice cream flavor.}$$

may be a statement for a specific person named Jon. The sentence

$$\text{Socrates was a good philosopher.}$$

is a less clear example. We should ask what constitutes a *good* philosopher; the sentence about the famous ancient Greek philosopher, Socrates, is a logical statement if we can give precise criteria for the adjective *good*. However, social commentary is not mathematics.

The examples given above are usually called *simple* statements since they cannot be broken into pieces that are themselves statements. Some authors call simple statements *atoms* or *atomic* statements.

You can probably think of examples that are not so simple, such as *compound* statements involving words like *not, and, or, if, then, implies*, and *equivalent*. We will use these terms, called *connectives*, in Section 2.2. The same authors who call simple statements either atoms or atomic statements call compound statements *molecules* or *molecular* statements.

You might also imagine statements involving variables. The sentences

$$\text{The square of every real number is nonnegative.}$$

and

> There exists a real number whose square is not positive.

are true statements, but they are different from our simple statements. Both contain variables. The former states a fact about *every* real number and the latter states a fact about *at least one* real number (specifically, 0). *Predicates*—sentences involving variables—that are *quantified*, as in the examples above, will be studied in Section 2.5.

You may say that you know exactly what the words *not, and, or,* etc. mean and how their use will change the meaning of a statement. However, the colloquial usage of these terms does not entirely agree with the mathematical usage. For this reason, we will define these words in terms of truth values—the truth values of compound statements formed by connectives are functions of the truth values of the simple statements from which they are formed.

Linguistics also comes into play. Translating between mathematical, or logical, language and a natural language, such as English, is not always straightforward. We will start by using statements in *symbolic logic* and their English translations, side by side. After this chapter, we will abandon the logical notation completely; however, we recommend that you adopt some of it as shorthand for your own writing of mathematics. We will remind you of this in Remark 43 at the end of this chapter.

2.2 Mathematical Connectives

We are now ready to study the different connectives used in creating compound statements. These are being defined in a formal way, based on truth values, rather than linguistically, based on the colloquial usage of English.

Negation (not)

An integer is called an *even number* when it equals twice some integer and is called an *odd number* when it equals one plus twice some integer. The statement

> 3 is an odd number.

is a true mathematical statement; the statement

> 3 is not an odd number.

is a false mathematical statement. How are these two statements related? You might say that the second is the negation of the first. This is correct. However, we will define negation not colloquially, but with truth values.

Definition 2. For a mathematical statement p, the *negation* of p is a mathematical statement, denoted $\neg p$, whose truth value is the opposite of the truth value of p.

We denote the negation of a statement, p, by *not p*; some denote it symbolically as either $\sim p$ or $-p$, rather than $\neg p$. We can illustrate Definition 2 using the following truth table:

p	$\neg p$
T	F
F	T

In a truth table, we read the truth values—T for true and F for false—of the column headings across a row. In the truth table above for *not p*, the first row indicates that whenever p is true, then *not p* is false while the second row indicates that whenever p is false, then *not p* is true. Truth tables give us a shorthand for expressing the truth values of related statements, starting with simple statements like p.

Colloquially, we generally negate a statement by using the word *not*. Of course, it isn't—sorry for the contraction, but we didn't want the word *not* to appear here—as simple as changing p to *not p*. For example, suppose p is the statement

3 is an odd number.

For *not p*, we write

3 is not an odd number.

rather than

Not 3 is an odd number.

Conjunction (and)

Two mathematical statements can be combined to form a more complex statement. One way to do that is to connect them with the word *and*. Let us define this connective by a truth table.

Definition 3. For mathematical statements p and q, the *conjunction* of p and q is the mathematical statement, denoted $p \wedge q$, whose truth value is false unless both p and q are true.

The truth value of a conjunction varies according to the following truth table:

p	q	$p \wedge q$
T	T	T
T	F	F
F	T	F
F	F	F

We will generally write *p and q* for the conjunction $p \wedge q$. Some authors use the notation $p \cdot q$ or $p \& q$.

An example of conjunction is the statement

2 is positive and 3 is negative.

It is false since

3 is negative

is a false statement.

Remark 4. The conjunction of two statements is a true statement exactly when both statements are true and is a false statement when at least one statement is false. It is sometimes

2.2. Mathematical Connectives

tempting to think that the conjunction "false and false" is a true statement, but this is nonsense, in everyday language as well as in mathematics.

Disjunction (or)

You might think we were being pedantic in defining negation and conjunction of mathematical statements. The next two definitions may change your mind. Is the statement

$$2 \text{ is positive or } 3 \text{ is positive}$$

true or false? Do not say this is false because it is not true that only one or the other is true; the statement is true. For a less mathematical example, would you conclude from

I took a bath or I took a shower

that the speaker took both a bath and a shower? Natural languages such as English are frequently ambiguous concerning the meaning of the word "or." It may mean one or the other or both. However, it may mean one or the other, but not both. The following definition resolves this ambiguity for mathematics.

Definition 5. For mathematical statements p and q, the *disjunction* of p and q is a mathematical statement, denoted $p \vee q$, whose truth value is true unless both p and q are false.

We see that the of truth value of a disjunction varies according to the following truth table:

p	q	$p \vee q$
T	T	T
T	F	T
F	T	T
F	F	F

We will generally write *p or q* for the disjunction $p \vee q$. As we said above, you should be extremely careful, since the word *or* is not used with its most common meaning. When someone says

I will give my ticket to Alex or to Jamie.

it is implied that only one person will receive the ticket: either Alex or Jamie, but not both. In mathematics, we use *or* in the sense of *at least one*.

Remark 6. The disjunction of two statements is true if at least one of the statements is true. In mathematics, the word *or* is equivalent to the hideous word *and/or*. As we use it in mathematics, *or* is referred to as the *mathematical or* and the *inclusive or* (since the possibility of both is included). Similarly, the *or* of everyday language is referred to as the *exclusive or* (since the possibility of both is excluded). Some people go to the extreme of using the new word *xor* for the *exclusive or*. We do not like it and you may (or may not) like it. Is *xor* any better (or worse) than *and/or*?

Example 7. We usually recognize negation, conjunction, and disjunction from the use of keywords: the word *not* in negations, the word *and* in conjunctions, and the word *or* in disjunctions. However, notation can hide these constructions. In the three true statements

$$1 \leq 1$$
$$2 < 3 < 4$$
$$5 \leq 6 \leq 6$$

the first is a disjunction because it can be interpreted as $1 < 1$ or $1 = 1$, the second is the conjunction $2 < 3$ and $3 < 4$, and the third is a conjunction of two disjunctions.

Implication (implies/if, then)

We now introduce two additional types of compound statements: implications and equivalences. These are also called *conditionals* and *biconditionals*, respectively, by some authors. The next construct, implication, also disagrees with the common colloquial interpretation.

Definition 8. For mathematical statements p and q, the *implication* (or *conditional* statement) that p *implies* q, denoted $p \Rightarrow q$, is a mathematical statement whose truth value is true unless p is true and q is false.

We see that the of truth value of an implication varies according to the following truth table:

p	q	$p \Rightarrow q$
T	T	T
T	F	F
F	T	T
F	F	T

We will generally write either *if p, then q* or *p implies q* for the implication $p \Rightarrow q$; some authors use the notation $p \rightarrow q$ or $p \supset q$ instead of $p \Rightarrow q$. For the implication p *implies* q, the statement p is called the *hypothesis* or *antecedent* and the statement q is called the *conclusion* or *consequent* of the implication.

Remark 9. In ordinary language, it is understood, for a statement of the form *if p, then q*, that there exists a causal connection between p and q, as in

If you leave me, then I will be badly hurt.

In mathematical language, a conditional statement is false only when the hypothesis is true and the conclusion is false. Since a false hypothesis always yields a true implication, no sense of causality should be inferred.

Example 10. Here are four examples:

If 2 is positive, then 3 is positive.

is a true statement since both the hypothesis and the conclusion are true statements.

If 2 is negative, then 3 is positive.

is a true statement since the hypothesis is false.

If 2 is negative, then 3 is negative.

2.2. Mathematical Connectives

is a true statement since the hypothesis is false.

$$\text{If 2 is positive, then 3 is negative.}$$

is a false statement since the hypothesis is true, but the conclusion is false.

Make sure you read the middle two examples carefully and understand why they are true! Do not let the trivial nature of the hypotheses and conclusions fool you: once the truth values of the hypothesis and the conclusion of any given implication are known, the truth value of that implication is trivially known as well.

Remark 11. An implication *if p, then q* is called *vacuously true* if its hypothesis, p, is false. For a vacuously true implication, it does not matter whether its conclusion is true or false. Look again at Example 10 and notice that we did not indicate whether the conclusion was true or false in the middle two examples since it did not matter.

We have already said that *if p, then q* and *p implies q* are used interchangeably to mean $p \Rightarrow q$. All of (1)–(6) are equivalent English sentences:

1. If p, then q.

2. p implies q.

3. p only if q.

4. q if p.

5. p is sufficient for q.

6. q is necessary for p.

Forms (3) and (4) may seem peculiar to you; think of them, for a true implication *p implies q*, as "p is true only if q is true" and "q is true if p is true." (You should compare these with the truth table for $p \Rightarrow q$.) The forms (5) and (6) are more esoteric. We have a prejudice against this language unless used together, as explained after the following example.

Example 12. The sentence

$$\text{For all real numbers } a \text{ and } b, \text{ if } a = 0, \text{ then } ab = 0.$$

is a statement of a different type since it contains variables. Notice that the truth value of "$a = 0$" depends on the value of the variable a. We will discuss such statements further in Section 2.5; we use it here because it better illustrates the different forms of implications mentioned above. The following forms should strike you as equivalent.

For all real numbers a and b, $a = 0$ only if $ab = 0$.

For all real numbers a and b, $ab = 0$ if $a = 0$.

For all real numbers a and b, $a = 0$ is sufficient for $ab = 0$.

For all real numbers a and b, $ab = 0$ is necessary for $a = 0$.

Equivalence (equivalent/if and only if/necessary and sufficient)

Our final construct, equivalence, probably agrees with your colloquial use of that word.

Definition 13. For mathematical statements p and q, the *equivalence* of p and q, denoted $p \Leftrightarrow q$, is a mathematical statement whose truth value is false unless p and q have the same truth value.

The truth value of an equivalence varies according to the following truth table:

p	q	$p \Leftrightarrow q$
T	T	T
T	F	F
F	T	F
F	F	T

Instead of writing $p \Leftrightarrow q$ for an equivalence, which is also called a *biconditional statement*, we will generally write one of the following:

p is equivalent to q,

p if and only if q,

p is necessary and sufficient for q.

Many people write p *iff* q in place of p *if and only if* q. Some authors use $p \leftrightarrow q$ or $p \equiv q$ instead of $p \Leftrightarrow q$.

Remark 14. In Section 1.3 of Chapter 1, we explained that definitions are often stated in the form of *if and only if* statements. Definitions are not logical statements whose truth value can be determined. You may choose to think of all mathematical terms as undefined, in which case definitions become axioms. This make the biconditional appearance of definitions more reasonable. Rather, for a definition in the form of an *if and only if* statement, the first half of the statement names the term being defined by the other half. In definitions, many authors simply write *if*. For example, we could define even integers as follows: an integer is *even* if it is two times some integer.

2.3 Symbolic Logic

We could make an entire course out of symbolic logic. Rather than do that, we will discuss some of the more important compound statements, those we will use occasionally. Before doing this, let us discuss how statements of symbolic logic are parsed. We will use the symbolic notation throughout this section and return to English statements in the next section.

Remark 15. A rule for the order of operations in arithmetic is that multiplications are performed before additions. So

$$1 + 2 \times 3 = 7 \neq 9.$$

There are also rules for the order of operations in logic. Reading from left to right, do things in the following order:

parentheses (or brackets, etc.)

2.3. Symbolic Logic

negation
conjunctions and disjunctions (in any order)
implication
equivalence

Groupings by parentheses, brackets, etc. are treated hierarchically. That is, nested groupings are evaluated from the inside towards the outside (just as for algebraic expressions). We think of this as working from the inside out.

Therefore, $\neg p \vee q$ is the same as $(\neg p) \vee q$ and different from $\neg(p \vee q)$. Between matched pairs of parentheses you must apply the same rules. For a more complicated example, the following two compound statements based on simple statements $a, b, c, \ldots, i$ are equivalent:

$$a \vee \neg b \wedge c \Rightarrow \neg d \wedge e \Leftrightarrow \neg f \wedge g \vee h \Rightarrow i,$$

$$\left\langle \left\{ [a \vee (\neg b)] \wedge c \right\} \Rightarrow [(\neg d) \wedge e] \right\rangle \Leftrightarrow \left\langle \left\{ [(\neg f) \wedge g] \vee h \right\} \Rightarrow i \right\rangle.$$

Notice that, in order to evaluate this, we must start with the innermost pairs of parentheses, then the square brackets, then the curly braces, and then, finally, the angle brackets.

Example 16. If p is true, then the disjunction $p \vee \neg p$ is true. (The fact that $\neg p$ is false in the case when p is true does not matter.) On the other hand, if p is false, then $\neg p$ is true and the disjunction $p \vee \neg p$ is true. This can be expressed by the following truth table:

p	$\neg p$	$p \vee \neg p$
T	F	T
F	T	T

The column for $p \vee \neg p$ has all T's in its column; that is, *p or not p* is always true. The moral of this example is that you should never ask your mathematics instructor a question like "Will this topic be covered on the exam or not?" because you are likely to receive the perfectly correct, yet uninformative, response "Yes."

Every simple statement, such as $1 = 1$ or $1 = 2$, is either true or false. If it is true, it is called a *tautology*. On the other hand, if it is false, it is called a *contradiction*. The more interesting situation is for compound statements.

Definition 17. A compound statement is a *tautology* if and only if it is true for all possible truth values of its component statements.

Tautologies are easily recognized in truth tables: their columns contains only T's. Next we examine the other extreme.

Example 18. If p is false, then the conjunction $p \wedge \neg p$ is false. (The fact that $\neg p$ is true in the case when p is false does not matter.) On the other hand, if p is true, then $\neg p$ is false and the conjunction $p \wedge \neg p$ is false. This can be expressed by the following truth table with only F's in the column for $p \wedge \neg p$.

p	$\neg p$	$p \wedge \neg p$
T	F	F
F	T	F

Definition 19. A compound statement is a *contradiction* if and only if it is false for all possible truth values of its component statements.

Contradictions are easily recognized in truth tables: their columns contains only F's. Contradictions are important in mathematics. In fact, we will see in the next chapter that contradictions can actually be useful in proving theorems.

Let us look at a couple of additional examples.

Example 20. The following truth table shows that the implication $p \Rightarrow p$ is a tautology.

p	$p \Rightarrow p$
T	T
F	T

Example 21. The equivalence $\neg\neg p \Leftrightarrow p$ is a tautology:

p	$\neg p$	$\neg\neg p$	$\neg\neg p \Leftrightarrow p$
T	F	T	T
F	T	F	T

The following shows that some data can be extraneous.

Exercise 22. Use truth tables to show that the following are tautologies:

(a) $p \wedge q \Rightarrow p$

(b) $p \Rightarrow p \vee q$

The following gives us an alternative definition of implication.

Exercise 23. Use a truth table to show that $p \Rightarrow q \Leftrightarrow \neg p \vee q$ is a tautology.

The following exercise may remind you of the concept of equality. Note these three properties of equality:

$$x = x,$$
$$x = y \text{ if and only if } y = x, \text{ and}$$
$$x = y \text{ and } y = z \text{ implies } x = z,$$

for all x, y, and z.

Exercise 24. Use truth tables to show that the following are tautologies:

(a) $p \Leftrightarrow p$

(b) $(p \Leftrightarrow q) \Leftrightarrow (q \Leftrightarrow p)$

(c) $(p \Leftrightarrow q) \wedge (q \Leftrightarrow r) \Rightarrow (p \Leftrightarrow r)$

Recall that $p \Rightarrow q$ can be read as *p only if q* and $q \Rightarrow p$ can be read as *p if q*. It seems reasonable that *p if and only if q* should be equivalent to *p if q and p only if q*. This suggests the following important result; we will use this often.

2.3. Symbolic Logic

Exercise 25. Use a truth table to show that
$$(p \Leftrightarrow q) \Leftrightarrow (p \Rightarrow q) \wedge (q \Rightarrow p)$$
is a tautology.

Remark 26. In Definition 13, we defined the equivalence of p and q, denoted by $p \Leftrightarrow q$. Note that Exercise 25 could have been used to define equivalence! That is, in Definition 13, we could have *defined* $p \Leftrightarrow q$ as $(p \Rightarrow q) \wedge (q \Rightarrow p)$ and *proved* the equivalence of $p \Leftrightarrow q$ with the definition given by the truth table in Definition 13.

We can use truth tables in this way to prove many such statements in logic. Keep in mind that this use of truth tables is just a shorthand. To write out the proof of Exercise 25 in words would go something like this:

Proof. There are four cases:

1. p is true and q is true,
2. p is true and q is false,
3. p is false and q is true, and
4. p is false and q is false.

First, suppose p is true and q is true. Hence, *if p, then q* is true and *if q, then p* is true. Therefore, the conjunction *if p, then q, and if q, then p* is true. Moreover, since p and q are true, *p is equivalent to q* is true. This completes the first case.

Second, ... □

Exercise 27. Complete the proof, started above, of Exercise 25 using words.

Let us return to conditional statements. Given statements p and q, we can make implications *p implies q* as well as *q implies p*. We can also take the negations of p and q and form other implications. Three of these constructions are important enough to have names.

Definition 28. The *converse* of the implication $p \Rightarrow q$ is the implication $q \Rightarrow p$.

That is, the converse of the statement if p, then q is the statement if q, then p. If a statement is true, then its converse may be either true or false; you can examine the possibilities in the following exercise.

Exercise 29. If possible, give examples of each of the following:

(a) a true implication whose converse is true,

(b) a true implication whose converse is false,

(c) a false implication whose converse is true,

(d) a false implication whose converse is false.

Let p and q be simple statements. The implication if p, then q can only be false when p is true and q is false. Similarly, if q, then p can only be false when q is true and p is false.

So, as you should have discovered in the previous exercise, it is impossible for both such an implication—between simple statements—and its converse to be false. If you thought otherwise, you probably introduced a variable in your statements. Soon, we will introduce variables and quantifiers; this will make things more difficult, but far more interesting.

Definition 30. The *contrapositive* of the implication $p \Rightarrow q$ is the implication $\neg q \Rightarrow \neg p$.

So, the contrapositive of the statement *if p, then q* is the statement *if not q, then not p*.

Example 31. The contrapositive of the statement

If the moon is made of cheese, then $1 = 2$

is the statement

If $1 \neq 2$, then the moon is not made of cheese.

You should recognize that both of these statements are true.

The truth value of a contrapositive statement is always the same as the truth value of the original statement.

> **Exercise 32.** Prove that an implication is equivalent to its contrapositive. That is,
> $$p \Rightarrow q \Leftrightarrow \neg q \Rightarrow \neg p.$$

Definition 33. The *inverse* of the implication $p \Rightarrow q$ is the implication $\neg p \Rightarrow \neg q$.

The inverse of the statement *if p, then q* is the statement *if not p, then not q*; it is the contrapositive of the converse of *if p, then q*. Therefore, the converse and the inverse of a given statement are equivalent.

> **Exercise 34.** Use a truth table to show that, given an implication, its converse and its inverse are equivalent; i.e., $q \Rightarrow p \Leftrightarrow \neg p \Rightarrow \neg q$.

> **Exercise 35.** Are an implication and its inverse equivalent? Prove your answer.

From every implication we can derive three others: the converse, the contrapositive, and the inverse.

$$\begin{array}{lrcl}
\text{Original:} & p & \Longrightarrow & q, \quad \text{if } p, \text{ then } q \\
\text{Converse:} & q & \Longrightarrow & p, \quad \text{if } q, \text{ then } p \\
\text{Contrapositive:} & \neg q & \Longrightarrow & \neg p, \quad \text{if not } q, \text{ then not } p \\
\text{Inverse:} & \neg p & \Longrightarrow & \neg q, \quad \text{if not } p, \text{ then not } q \\
\end{array}$$

2.4 Compound Statements in English

Statements can get complex and writing statements in English is further complicated because we do not use parentheses to indicate order of thoughts; commas can play an important role. We use italics and quotation marks to try to clarify sentences in this chapter, but this is a luxury upon which we should not depend.

In
$$1 \text{ is negative and } 2 \text{ is negative or } 3 \text{ is positive}$$
if we let p denote the statement 1 is negative, let q denote the statement 2 is negative and let r denote the statement 3 is positive, then our compound statement looks like

$$p \wedge q \vee r.$$

For symbolic logic, the order of operations dictates that this is equivalent to

$$(p \wedge q) \vee r.$$

Since p and q are false and r is true, this statement is true. Notice that $p \wedge q \vee r$ is different from $p \wedge (q \vee r)$; in fact, $p \wedge (q \vee r)$ is false.

How do we make this distinction in written English? In the following two statements

$$1 \text{ is negative and } 2 \text{ is negative, or } 3 \text{ is positive.}$$
$$1 \text{ is negative, and } 2 \text{ is negative or } 3 \text{ is positive.}$$

there certainly is a lot of weight riding on those two little commas. The former should be interpreted as $(p \wedge q) \vee r$ while the latter should be interpreted as $p \wedge (q \vee r)$.

In spoken English, it is very difficult to hear those little commas!

An alternative is to rephrase the sentences. This is not so simple. For example,

The conjunction of the statement 1 is negative and the

disjunction of the statements 2 is negative and 3 is positive.

While this may be clearer, it is not a satisfying solution. Look at the symbolic statement in Exercise 25. In words, we could say something like the following:

p if and only if q is equivalent to the conjunction

of the implications if p, then q and if q, then p.

2.5 Predicates and Quantifiers

Take another look at Example 12. Also, did you have any difficulties with Exercise 29? These were tricky since they got you thinking (perhaps, in the case of Exercise 29) about introducing variables into statements. In this section, we introduce the use of variables in both logical and mathematical statements. The statements of most mathematical theorems involve variables. By the way, variables are sometimes called *free* variables, to emphasize that they are free to change.

The statement
$$\text{The square of } -2 \text{ is positive.}$$
is true since $(-2)^2 = 4$.

Now, the sentence

$$\text{The number } x \text{ is positive.}$$

is not a logical statement since an unresolved variable is involved. No answer can be given as to the truth value of the sentence unless x is given a specific value. For example, if $x = 2$, the sentence becomes a true statement, but if $x = -4$, the sentence becomes a false statement.

Any declarative sentence that contains one or more unresolved variables is called a *predicate* if and only if it is a statement whenever specific values are substituted for the variables. The word *unresolved* means that no values, or possible set of values, are given for the variables. We will return to the business of resolving, or substituting for, the variables shortly. In mathematical logic, a predicate is called a *propositional function* or a *sentential function*.

Two other examples of predicates are $x \geq 0$ and $x + y = 0$; the former has one variable and the latter has two variables. Using functional notation, we could denote these predicates by $P(x)$ and $Q(x, y)$, respectively.

Just as in the case of logical statements, we can negate a predicate. For instance, if $P(x)$ is the predicate $x \geq 0$, then *not* $P(x)$ is the predicate $x \not\geq 0$. Thinking of the real numbers, we could have said $x < 0$, but, technically, that requires proof!

Predicates also can be combined to form new predicates as is illustrated in the following exercise.

Exercise 36. Equations and inequalities involving variables are predicates.
(a) Determine all real numbers x for which the conjunction

$$x + 3 > 6 \text{ and } x^2 < 3$$

is a true statement.
(b) Similarly, determine all real numbers x and y for which the disjunction

$$xy = 0 \text{ or } 2x - 3y = 0$$

is a true statement.

Substituting a single value for a variable is not generally of interest since it can be done directly, thereby turning a predicate into a statement. For example, instead of writing a predicate with a substitution such as

$$x^2 > 3 \text{ for } x = 2.$$

we could just write the statement

$$2^2 > 3.$$

However, there are two important ways of creating statements out of predicates; this is called *quantifying* a predicate.

The sentence

$$\text{For all real numbers, } x^2 = 4.$$

is a false statement since $5^2 \neq 4$. This is one kind of quantifier.

2.5. Predicates and Quantifiers

Definition 37. Let S be a set and let $P(x)$ be a predicate. The *universal quantifier* is a phrase of the form *for all $x \in S$*, denoted $\forall x \in S$. The *universally quantified predicate sentence*
$$\forall x \in S, P(x)$$
is a mathematical statement that is true if and only if $P(x)$ becomes a true statement whenever an element of S is substituted for x in the predicate $P(x)$.

We will generally write *for all $x \in S$, $P(x)$* for $\forall x \in S, P(x)$. The phrases *for every $x \in S$* and *for any $x \in S$* are also used for universal quantifiers. We recommend that you be careful with the use of the word *any* since it may be misinterpreted as *some* rather than *every*. The phrase *for all real numbers x* is equivalent to *for every $x \in \mathbb{R}$*.

Example 38. Which of the following statements are true and which are false?
(a) For all real numbers x, $(x+3)^2 + (x-2)^2 > 0$.
(b) For all real numbers x, $\sqrt{x^2} = x$.

For (a), the squares are nonnegative. $(x+3)^2 = 0$ only if $x = -3$ and $(x-2)^2 = 0$ only if $x = 2$. So they cannot both be equal to 0 for the same x. Therefore, the statement in (a) is true.

For (b), the square root notation denotes the nonnegative square root. For $x = -1$,
$$\sqrt{(-1)^2} = \sqrt{1} = 1 \neq -1.$$
Therefore, statement (b) is false.

Remark 39. Sometimes a universal quantifier is hidden in a statement. It is sometimes preferable, for purposes of logical manipulation, to restate such a statement, as is done in the next example. Sometimes the set defining the universal quantifier is hidden. In the statement

For all x, there exists a positive square root.

the set is not stated. This statement is true for the set of positive real numbers, but is false for the set of real numbers. When the set is omitted, we will assume that it is the set of real numbers.

Example 40. In geometry, the isosceles triangle theorem may be stated:

The angles at the base of an isosceles triangle are congruent.

This could be restated as follows:

For every triangle T, if T is isosceles,

then the angles at the base are congruent.

You might even wonder exactly what set of triangles we are talking about!

The sentence

There exists x, such that $x^2 = 4$.

is a true statement since $2^2 = 4$; it is also true since $(-2)^2 = 4$. The expression "there exists" is a new type of quantifier.

Definition 41. Let S be a set and let $P(x)$ be a predicate. The *existential quantifier* is a phrase of the form *there exists $x \in S$*, denoted $\exists x \in S$. The *existentially quantified predicate* sentence
$$\exists x \in S, P(x)$$
is a mathematical statement that is true if and only if $P(x)$ becomes a true statement for at least one element of S substituted for x in the predicate $P(x)$.

We will generally write *there exists $x \in S$ such that $P(x)$* for $\exists x \in S, P(x)$. The phrase *for some $x \in S$* is also used for existential quantifiers. Again, we remind you to be careful with the use of the word *any*.

Example 42. Which of the following statements about sets of real numbers are true and which are false?

(a) There exists x such that $(x + 5)^2 = 0$.

(b) There exists x such that $|x - 1| = x$.

(c) There exist x, y such that $x^2 + y^2 = 1$.

(d) There exists x such that $\sin^2 x + \cos^2 x = 0$.

Statement (a) is true since $x = -5$ satisfies the equation. Statement (b) is true since $x = \frac{1}{2}$ satisfies the equation. Statement (c) is true since, for example, $x = y = \sqrt{2}/2$, satisfies the equation. Statement (d) is false since $\sin^2 x + \cos^2 x = 1$ for all x.

Notice that the set of values for x is important: if the word *real* were changed to *natural*, then all three statements above would be false.

Let us see how we negate a sentence that is introduced by a universal quantifier. The statement
$$\text{For all } x \text{ and } y, x + y = 0.$$
is false. What is the negation of this sentence? If not all pairs of numbers add up to zero, there exist numbers x and y, whose sum is not zero, so the negation of our original sentence is
$$\text{There exist } x \text{ and } y \text{ such that } x + y \neq 0.$$
This is a true statement. In general, the negation of the statement
$$\text{For all } x, P(x).$$
is the statement
$$\text{There exists an } x \text{ such that not } P(x).$$
The negation of predicates with more than one variable is defined in an analogous way.

To see how we negate a sentence that is introduced by the existential quantifier, we look at the following example. The statement
$$\text{There exists a number } x, \text{ such that } x^2 < 0.$$
is a false statement. If there does not exist a number whose square is negative, then the square of all numbers will be nonnegative, so the negation of the original sentence is
$$\text{For all numbers } x, x^2 \not< 0,$$

2.5. Predicates and Quantifiers

which is equivalent to
$$\text{For all numbers } x, x^2 \geq 0,$$

In general, the negation of the statement
$$\text{There exists } x \text{ such that } P(x).$$
is the statement
$$\text{For all } x, \text{ not } P(x).$$

Remark 43 (logical notation). Although we have mentioned the symbols $\neg$, $\wedge$, $\vee$, $\Rightarrow$, $\Leftrightarrow$, $\forall$, and $\exists$, we will not use them after this chapter. Many people use them in writing mathematics by hand. The symbols for implications, equivalences, and quantifiers are especially useful. Some people also use $\ni$ or *s.t.* for "such that," $\because$ for "since" or "because," and $\therefore$ for "therefore" or "thus" or "hence."

We prefer to use double barred arrows ($\Rightarrow$, $\Leftarrow$, $\Leftrightarrow$) for logical operations since single barred arrows, especially $\rightarrow$, are used frequently in the notation for functions.

We recommend that you use whatever shorthand you feel comfortable with. Moreover, we want to remind you to keep in mind your audience, those who will read what you write!

2.6 Supplemental Exercises

Definition Review. There were sixteen italicized terms defined in this chapter. You can take the truth values true and false to be undefined. Let p and q be logical or mathematical statements (or predicates). Define each of the following and give an example of each:

Definition 1. p is a *logical statement*

Definition 1. p is a *mathematical statement*

Definition 2. $\neg p$ is the *negation* of a statement p

Definition 3. $p \wedge q$ (p *and* q) is the *conjunction* of statement p and q

Definition 5. $p \vee q$ (p *or* q) is the *disjunction* of statement p and q

Definition 8. the *implication* $p \Rightarrow q$ (p *implies* q or *if* p, *then* q)

Definition 13. the *equivalence* $p \Leftrightarrow q$ of statements p and q

Definition 17. a compound statement is a *tautology*

Definition 19. a compound statement is a *contradiction*

Definition 28. the *converse* of an implication $p \Rightarrow q$

Definition 30. the *contrapositive* of an implication $p \Rightarrow q$

Definition 33. the *inverse* of an implication $p \Rightarrow q$

Definition 37. the *universal quantifier*

Definition 37. an *universally quantified predicate* sentence

Definition 41. the *existential quantifier*

Definition 41. an *existentially quantified predicate* sentence

---

Supplemental Exercise 1. Are the following statements true or false? Explain.
   (a) $1 = 2$ or $3 = 3$.
   (b) $1 = 2$ and $3 = 3$.
   (c) $1 = 1$ or $3 = 3$.
   (d) $1 = 1$ and $3 = 3$.

Supplemental Exercise 2. Are the following statements true or false? Explain.
   (a) If $1 = 2$, then $3 = 3$.
   (b) If $1 = 2$, then $3 = 4$.
   (c) If $1 = 1$, then $3 = 3$.
   (d) If $1 = 1$, then $3 = 4$.

Supplemental Exercise 3. Are the following statements true or false? Explain.
   (a) For all real numbers x, $x > 0$.
   (b) There exists a real number x such that $x > 0$.

Supplemental Exercise 4. Are the following statement true or false? Explain.
   (a) For all real numbers x, $x \geq 0$ or $x \leq 0$.
   (b) There exists a real number x such that $x \geq 0$ and $x \leq 0$.

2.6. Supplemental Exercises

In Supplemental Exercises 5–38, use truth tables to show that each statement is a tautology.

Supplemental Exercise 5. $p \wedge (p \Rightarrow q) \Rightarrow q$

Supplemental Exercise 6. $\neg p \wedge (q \Rightarrow p) \Rightarrow \neg q$

Supplemental Exercise 7. $\neg p \wedge (p \vee q) \Rightarrow q$

Supplemental Exercise 8. $p \Rightarrow (q \Rightarrow p \wedge q)$

Supplemental Exercise 9. $(p \Rightarrow q) \wedge (q \Rightarrow r) \Rightarrow (p \Rightarrow r)$

Supplemental Exercise 10. $(p \wedge q \Rightarrow r) \Leftrightarrow (p \Rightarrow [q \Rightarrow r])$

Supplemental Exercise 11. $(p \wedge q \Rightarrow r) \Leftrightarrow (p \Rightarrow [q \Rightarrow r])$

Supplemental Exercise 12. $(p \Rightarrow [q \wedge \neg q]) \Rightarrow \neg p$

Supplemental Exercise 13. $(p \Rightarrow q) \Rightarrow (p \vee r \Rightarrow q \vee r)$

Supplemental Exercise 14. $(p \Rightarrow q) \Rightarrow ([q \Rightarrow r] \Rightarrow [p \Rightarrow r])$

Supplemental Exercise 15. $(p \Leftrightarrow q) \Leftrightarrow (q \Leftrightarrow p)$

Supplemental Exercise 16. $(p \Rightarrow q) \wedge (r \Rightarrow q) \Leftrightarrow (p \vee r \Rightarrow q)$

Supplemental Exercise 17. $(p \Rightarrow q) \wedge (p \Rightarrow r) \Leftrightarrow (p \Rightarrow q \wedge r)$

Supplemental Exercise 18. $(p \vee q) \Leftrightarrow (q \vee p)$

Supplemental Exercise 19. $(p \wedge q) \Leftrightarrow (q \wedge p)$

Supplemental Exercise 20. $(p \vee q) \vee r \Leftrightarrow p \vee (q \vee p)$

Supplemental Exercise 21. $(p \wedge q) \wedge r \Leftrightarrow p \wedge (q \wedge p)$

Supplemental Exercise 22. $p \vee (q \wedge r) \Leftrightarrow (p \vee q) \wedge (p \vee r)$

Supplemental Exercise 23. $p \wedge (q \vee r) \Leftrightarrow (p \wedge q) \vee (p \wedge r)$

Supplemental Exercise 24. $p \vee p \Leftrightarrow p$

Supplemental Exercise 25. $p \wedge p \Leftrightarrow p$

Supplemental Exercise 26. $\neg(p \vee q) \Leftrightarrow \neg p \wedge \neg q$

Supplemental Exercise 27. $\neg(p \wedge q) \Leftrightarrow \neg p \vee \neg q$

Supplemental Exercise 28. $p \Rightarrow q \Leftrightarrow \neg p \vee q$ (This gives us a useful equivalent definition of implication.) This implies: $\neg(p \Rightarrow q) \Leftrightarrow p \wedge \neg q$.

Supplemental Exercise 29. $p \Rightarrow q \Leftrightarrow \neg(p \wedge \neg q)$

Supplemental Exercise 30. $p \vee q \Leftrightarrow \neg p \Rightarrow q$

Supplemental Exercise 31. $p \vee q \Leftrightarrow \neg(\neg p \wedge \neg q)$

Supplemental Exercise 32. $p \wedge q \Leftrightarrow \neg(p \Rightarrow \neg q)$

Supplemental Exercise 33. $p \vee q \Rightarrow r \Leftrightarrow (p \Rightarrow r) \wedge (q \Rightarrow r)$

Supplemental Exercise 34. $p \wedge q \Leftrightarrow \neg(\neg p \vee \neg q)$

In Supplemental Exercises 35–38, let F and T represent statements that are always false or true, respectively.

Supplemental Exercise 35. $p \vee F \Leftrightarrow p$

Supplemental Exercise 36. $p \wedge F \Leftrightarrow F$

Supplemental Exercise 37. $p \vee T \Leftrightarrow T$

Supplemental Exercise 38. $p \wedge T \Leftrightarrow p$

Supplemental Exercise 39. Is the following statement true or false? Explain.
    For all x and y, $|x - y| = |y - x|$.

Supplemental Exercise 40. Show that $q \wedge (p \Rightarrow q) \Rightarrow p$ is not a tautology. (This is sometimes referred to as the fallacy of asserting the conclusion.)

Supplemental Exercise 41. Show that $p \wedge \neg q \Rightarrow F \Leftrightarrow p \Rightarrow q$ is a tautology. (This is sometimes referred to as proof by contradiction (you will use this idea in Section 3.3 of Chapter 3. In the statement, F denotes a false statement.)

Supplemental Exercise 42. A statement can have more than one quantifier. While the order of quantifiers of the same type does not matter, the order of mixed quantifiers does matter. Explain why
$$\exists x \in S \; \forall y \in S \; x = y$$
and
$$\forall x \in S \; \exists y \in S \; x = y$$
are not equivalent when $S = \mathbb{R}$, but are equivalent when $S = \{0\}$.

Supplemental Exercise 43. Project. Investigate the role of formal logic in the development of mathematical practice and philosophy.

3
Proofs in Mathematics

3.1 What is Mathematics?

Using the ideas of the previous section, we are now in a position where we can understand the formal nature of a given statement. The next important step would be to provide a justification for the truth or falsity of this statement. Since the invention of the deductive method in Ancient Greece, proof has been the means of warranting mathematical knowledge. But justification is not the only function of proof. Proof systematizes various mathematical results, explains why they are true, and enhances mathematical understanding and communication. In addition, it promotes the development of mathematical competence and the discovery of new mathematical facts.

Mathematics is, foremost, a creative activity. Theorems are not handed out by God (or professors) to the faithful (students?) with the expectation that the supplicants (students?) will supply the appropriate proof before awaiting the next theorem from on high. (Some philosophers disagree with this point of view.) Theorems are a product of human endeavor, dependent on the human experience and knowledge. To do mathematics is to create mathematics: theorems along with their proofs. Yes, perhaps the scholastic experience is slightly different. Many professors and their texts play the roles of gods and kings, expecting their slaves (students!) to do the grungy work.

This text deals with constructing proofs, not discovering theorems. From elementary school through the master's degree, students of mathematics rarely encounter the other side, the creation of the theorems themselves. Finding theorems is the primary goal of a doctoral dissertation and of mathematical research.

Most times you are given a proof immediately after the theorem. Your job is then to comprehend both the theorem and its proof. In order to learn how to do proofs, you should think about how you would prove the theorem first. To start thinking like a mathematician you must read slowly, pausing to think often. If you pause now and then to ask the question "What can I try to prove next?" you will be better able to construct proofs and perhaps you will be taking your first firm step towards becoming a mathematician.

Two things that you should do frequently, which are helpful both in finding theorems and proofs (not surprising since they go hand in hand), are to think of examples and draw pictures. In fact, these are not unrelated activities. Often, working out an example will help you to better understand what makes a theorem true, that is, how to prove it. Pictures are usually an extension of that process. Change the hypotheses and consider examples where the conclusion fails. These counterexamples can help you to isolate the missing ideas that make up the proof.

You may be thinking that you have seen lots of pictures in mathematics texts that you did not find the least bit illuminating. That is understandable. The important part of the picture is usually in its construction, not in its final presentation. It is like looking up the answer to an exercise in the back of a text. The answer alone does not illuminate the path from the problem to the answer.

If you get nothing else out of this chapter, we hope you will understand two things. First, you must spend far more time thinking than reading. (It is not uncommon for a mathematician to take hours, even days, to read and comprehend a single page!) Second, you should always think of examples (and counterexamples), perhaps drawing pictures as a guide. (Yes, you can draw mental images, but it is surprising how different things can look on paper.)

In the remainder of this chapter, we will examine various techniques of proof.

3.2 Direct Proof

A mathematical statement of the form

$$\text{for all } x \in S, \text{ if } p(x), \text{ then } q(x).$$

is a universally quantified predicate statement where $p(x)$ and $q(x)$ are predicates. In mathematics, the most common statements that will require a proof are of this form.

Here is the last truth table that you will see in this text. It describes perfectly what needs to be done: to prove that the statement given above is a tautology, we must show that for every x such that $p(x)$ is true, $q(x)$ is also true.

$p(x)$	$q(x)$	$p(x) \Rightarrow q(x)$
T	T	T
T	F	F
F	T	T
F	F	T

Looking at the truth table, the proof amounts to showing that the situation of the second row (the one where the implication is false) never happens! (When the statement "for all x $p(x) \Rightarrow q(x)$" is true because $p(x)$ is always false, we say the statement is *vacuously true*. Compare this with Remark 11 of Chapter 2.)

The first type of proof we will consider is direct proof. When someone says that a proof is "by brute force" or by "following your nose," they are referring to a direct proof. While "brute force" implies some sort of explicit computation, all the descriptions describe the basic nature of a direct proof.

Remark 1. Many mathematicians will say that they will "construct" a proof. This naive use of the word "construct" may be misunderstood. ("Constructive proof" may refer to the constructivist philosophy of mathematics. Not everyone subscribes to this philosophy. In this philosophy, the existence of a mathematical object is accepted only when it can be constructed, for instance, only when an explicit example is known). This may or may not indicate a direct proof.

3.2. Direct Proof

The basic idea of direct proof is simple: start from the hypotheses, make a sequence of implications, and arrive at the conclusion.

An Example of a Direct Proof

Here are the first five axioms from Euclid's *Elements*. This book, written around 300 B.C., was the prototype of mathematical exposition for two millennia.

Axiom 1. If two straight lines intersected by a transversal make the sum of the measures of the interior angles on the same side less than 180°, then the two straight lines meet on that same side.

Axiom 2. Things that are equal to the same thing are equal to one another.

Axiom 3. If equals are added to the same thing, the sums are equal.

Axiom 4. If equals are subtracted from equals, the differences are equal.

Axiom 5. All right angles are equal to one another.

These axioms are referred to in the next proof.

Proposition 2. *If two different lines intersect, then their vertical angles are equal.*

Before writing down a proof, let us start with a picture. Consider Figure 3.1, which shows two intersecting lines.

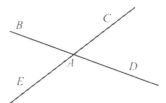

Figure 3.1. Vertical angles

If nothing else, the figure gives us a shorthand notation for our proof by giving names to the angles. By the way, if you do not know the definition of vertical angles, you should be able to figure it out from the proposition and the figure! We wish to show that $\angle BAC$ equals $\angle DAE$ and that $\angle CAD$ equals $\angle EAB$.

Proof. Let BAD and CAE be two lines intersecting at the point A. Then the angles $\angle BAD$ and $\angle CAE$ are both equal to two right angles. By Axiom 5 and the familiar construction of perpendicular lines, the angle at a point on a line is equal to two right angles. In more modern language, we can say these angles measure π or 180°. If we subtract from both angles the angle $\angle CAD$, by Euclid's Axiom 4, we conclude that $\angle BAC = \angle DAE$, which is what we wanted to prove.

Using the same line of reasoning, we can prove that $\angle CAD = \angle EAB$. □

Exercise 3. Complete the proof of Proposition 2: prove $\angle CAD = \angle EAB$.

If you prefer to think about the measures of angles, then you might rewrite the above proof as follows:

Proof. Let BAD and CAE be two lines intersecting at the point A. Then

$$m\angle BAD = \pi \quad \text{and} \quad m\angle CAE = \pi.$$

Therefore, $m\angle BAD = m\angle CAE$ and

$$m\angle BAD - m\angle CAD = m\angle CAE - m\angle CAD.$$

Since

$$m\angle BAC = m\angle BAD - m\angle CAD$$

and

$$m\angle DAE = m\angle CAE - m\angle CAD,$$

we have proved that $m\angle BAC = m\angle DAE$. Using the same line of reasoning, you can prove that $m\angle CAD = m\angle EAB$. □

Remark 4. It is always good to draw pictures, where practical, to help your thought processes along. Be careful that you do not make assumptions that are derived from your sketches. For example, when proving something about two intersecting lines, do not have them intersect at right angles.

Proof by Brute Force: Another Example of Direct Proof

By definition, every even integer can be expressed in the form $2n$ for some integer n. Also, every odd integer can be expressed in the form $2n + 1$ for some integer n. We prove several statements regarding even and odd integers.

Proposition 5. *The sum of two even integers is even.*

Proof. Start with any two even integers, x and y. Since x and y are even, there exist integers m and n such that $x = 2m$ and $y = 2n$. Then

$$x + y = 2m + 2n = 2(m + n) = 2k,$$

where k is the integer $m + n$. Hence, $x + y$ is even. □

Remark 6. In the proof of Proposition 5, we have tacitly assumed that the sum of two integers is an integer. This can be proved. In fact, the construction of the integers and of addition accomplish this. It is surprisingly difficult; see Chapter 10.

Proposition 7. *The square of any odd integer is odd.*

Proof. Every odd integer is of the form $x = 2n + 1$ for some integer n. Then

$$x^2 = (2n + 1)^2 = 4n^2 + 4n + 1 = 2(2n^2 + 2n) + 1 = 2k + 1,$$

where k is the integer $2n^2 + 2n$. Hence, x^2 is odd. □

Exercise 8. Prove that the square of an even integer is even.

Proof by cases

In proving a proposition, we sometimes have to distinguish cases. The following example is suggestive.

Example 9. To prove that $(-1)^n + (-1)^{n+1} = 0$ for all integers n, we have two cases: n is either even or odd.

Case 1: suppose n is even. Then $n+1$ is odd. (Why is this true? You should prove this from our definitions of even and odd.) Hence, $(-1)^n = 1$ and $(-1)^{n+1} = -1$. Therefore, the sum equals 0.

Case 2: suppose n is odd. Then $n+1$ is even. Hence, $(-1)^n = -1$ and $(-1)^{n+1} = 1$. Therefore, the sum equals 0.

We give another proof of $(-1)^n + (-1)^{n+1} = 0$ for all integers n:

$$\begin{aligned}(-1)^n + (-1)^{n+1} &= (-1)^n + (-1)^n \cdot (-1) \\ &= (-1)^n \cdot (1 + (-1)) \\ &= 0.\end{aligned}$$

The moral of this example is that there is often more than one way to prove a theorem. The usual problem is to find *some* proof. Some mathematicians like to find "the best" proof of every theorem; of course, this is subjective. The twentieth century mathematician P. Erdős liked to say that God has *The Book,* which contains all of the theorems, along with their best proofs.

You may recall the definition of the absolute value, $|x|$, of a real number, x. We can define $|x|$ to equal $\sqrt{x^2}$. However, the first definition you saw was probably something like

$$|x| = \begin{cases} x & \text{, if } x \geq 0 \\ -x & \text{, if } x < 0 \end{cases},$$

where $-x$ plays the role of "dropping the negative sign" when x is negative. Since this is a definition by cases, proofs of statements involving absolute values often split into cases.

Example 10. Let us prove that $|x| \geq x$ for all $x \in \mathbb{R}$. There are two cases according to the definition of absolute value. If $x \geq 0$, then $|x| = x$ by definition; hence $|x| \geq x$. On the other hand, if $x < 0$, then $|x| = -x$ is positive; hence $|x| \geq x$.

Exercise 11. Prove that for all $x \in \mathbb{R}$, $|2x + 1| > x$.

Remark 12. We would like to point out that "proof by cases" is not a proof technique at all. Instead, "proof by cases" is a method of reducing a problem to easier, smaller problems. In fact, the methods of proof used to prove each of the individual cases may be different. Compare with Supplemental Exercise 33 in Chapter 2.

3.3 Contraposition and Proof by Contradiction

An implication is equivalent to its contrapositive. Therefore, to prove a statement, we can prove its contrapositive. We call this proof by contraposition.

An Example of Proof by Contraposition

We prove the converse of Exercise 8.

Proposition 13. *If the square of an integer is even, then the integer is even.*

Proof. By contraposition, this statement is equivalent to saying "If an integer is odd, then its square is odd," which is Proposition 7. Since every statement is logically equivalent to its contrapositive, this completes the proof. □

The following exercises are examples of statements whose proofs are most easily done by contraposition.

Exercise 14. Prove that if $\frac{2n}{1+n^2}$ is irrational, then n is irrational.

Exercise 15. Prove that, for all real numbers a, if $a^2 > 0$, then $a \neq 0$.

The converse of Proposition 7 can be proved by contraposition. In the following exercise, we show that an equivalence can be proved by proving the "forwards" and "backwards" implications, as suggested by Exercise 25 of Chapter 2, which says the following:

$$(p \Leftrightarrow q) \Leftrightarrow (p \Rightarrow q) \wedge (q \Rightarrow p).$$

So, in the following exercise, it remains to prove that the square of an even integer is even.

Exercise 16. Prove that the square of an integer n is odd if and only if n is odd.

An idea similar to contraposition is proof by contradiction. In proof by contraposition, we used the fact that an implication and its contrapositive are logically equivalent. Mechanically, this amounts to the following: to prove that p implies q, we assume that the negation of q is true (i.e., q is false) and proceed to show that the negation of p is true (i.e., p is false). Seen as a direct proof, the hypothesis p has been contradicted since both p and its negation are apparently true. We use this idea to generalize this proof technique.

In proving an implication, p implies q, by contradiction, if we first suppose the hypotheses of a statement, p, and we assume the negation of its conclusion, q, and use these to derive a contradiction (to a hypothesis or to our assumption or to any other true statement), then we have proved the original implication.

Remark 17. Although the words "assume" and "suppose" are synonyms, it is helpful to use the word "assume" only when constructing a proof by contradiction. We "suppose" the hypotheses, but we "assume" the negation of the conclusion. In this way the word "assume" is a reminder that you are trying to find a contradiction.

Remark 18. The only difference in proofs by contraposition and by contradiction is in what true statement gets contradicted. The case of contradicting the hypothesis in a proof by contradiction is proof by contraposition. In a proof by contradiction, any true statement— including the hypotheses that are supposed and the negation of the conclusion that is assumed—may be contradicted.

3.3. Contraposition and Proof by Contradiction

Examples of Proof by Contradiction

Before we look at any examples of proof by contradiction, let us examine a faulty attempt at a proof.

Absurdum 19. *If the sum of two integers is even, then both integers are even.*

If a theorem deserves a proof, then an absurdum deserves a poof!

Poof? The "poof" is by contradiction. Suppose that x and y are integers and $x + y$ is an even integer. Negating the conclusion of the statement, we assume that x is even and y is odd. So, there exist integers m and n such that $x = 2m$ and $y = 2n + 1$. Then

$$x + y = 2m + (2n + 1) = 2(m + n) + 1$$

is an odd integer. This contradicts the fact that $x + y$ is even. Therefore, the assumption is false and we have that x and y are both even. □

> **Exercise 20.** Explain why the statement of Absurdum 19 is absurd and why its "poof" is wrong.

Here are two examples of proof by contradiction, for Propositions 21 and 22.

Proposition 21. *The sum of an odd integer and an even integer is odd.*

Proof. Let x be odd and y be an even integer. Assume that $x + y$ is not odd, that is, $x + y$ is even. Then $x = 2n + 1$ for some integer n, $y = 2m$ for some integer m and $x + y = 2p$ for some integer p. Hence, $x + y = 2n + 1 + 2m = 2p$. This is equivalent to $2(p - m - n) = 1$. But this is a contradiction since the number on the left-hand side is even, but 1 is odd. Therefore, $x + y$ is odd. □

This proof would not make Erdős's *The Book* (see Example 9). The direct proof is shorter and neater.

The second example of a proof by contradiction is given in Theorem 22. This is a famous theorem.

Theorem 22. $\sqrt{2}$ *is irrational.*

Before giving a proof of this proposition, you might wonder how we know that there exists a real number called $\sqrt{2}$. We can give an informal geometric argument. The existence of this number follows from the Pythagorean Theorem and the assertion that every line segment has a length. Just think of the length of the diagonal of a square whose side is 1. A more algebraic proof can be given by constructing the set of real numbers in which the irrational numbers are constructed from the rational numbers. (See Part III.) The following proof shows that this real number is irrational.

Proof of Theorem 22. Assume that $\sqrt{2}$ is rational. Then there exist natural numbers m and n such that

$$\sqrt{2} = \frac{m}{n}.$$

Without loss of generality, we assume that the fraction $\frac{m}{n}$ is reduced to lowest terms; hence, m and n are not both even. Squaring both sides we get

$$\frac{m^2}{n^2} = 2.$$

This is equivalent to the equality

$$2n^2 = m^2.$$

Since the left-hand side is an even number, m^2 is even. By Proposition 13, m is even. This means that

$$m = 2k$$

for some natural number k. Hence

$$2n^2 = (2k)^2 = 4k^2,$$

which implies that

$$n^2 = 2k^2.$$

Therefore, n^2 is even and so n is even. This is a contradiction since we assumed that m and n are not both even. Hence, $\sqrt{2}$ is irrational. □

Remark 23. The following remarks concern the style of writing proofs by contradiction.
(1) Many mathematicians start their proofs by indicating how they plan to proceed. A good way to start a proof by contradiction is by a statement such as "This proof is by contradiction."
(2) It is common to leave out the punch line, i.e., the declaration of a contradiction and the negation of the original assumption. Sometimes such proofs end with the simple declarative statement "Contradiction." This may leave the reader with a little bit to think about! For example, a much terser proof of Proposition 21 may look as follows.

Proof. Suppose $x = 2n + 1$ is odd and $y = 2m$ is even. Assume $x + y = 2p$ is even. Hence, $x + y = 2(n + m) + 1 = 2p$. Contradiction. □

We strongly suggest that you write as much as is necessary so that your fellow students could understand your proofs.

We will use the following important theorem to generalize Proposition 13 and Theorem 22; its proof is postponed until the Exercise 46. A natural number p is *prime* if and only if the only natural numbers that divide p are 1 and p. Note that 1 is not prime.

Theorem 24 (Fundamental Theorem of Arithmetic). *Each natural number other than 1 is the product of a uniquely determined finite collection of primes.*

For example,
$$12 = (2)(2)(3) = (2)(3)(2) = (3)(2)(2).$$

Each factorization includes two 2's and one 3. We think of 2 as a "product" of only one prime, namely 2. That is, every integer m greater than 1 can be written uniquely in the form

$$m = p_1^{n_1} p_2^{n_2} \cdots p_k^{n_k},$$

3.4. Proof by Induction

where
$$p_1 < p_2 < \cdots < p_k$$
are k distinct prime numbers and $n_1, n_2, \ldots, n_k$ are natural numbers.

Proposition 25 is a generalization of Proposition 13; when $n = p = 2$, the propositions are the same.

Proposition 25. *Let n be a natural number, m be an integer greater than 1, and p be a prime number. If m^n is divisible by p, then m is also divisible by p.*

Proof. By the Fundamental Theorem of Arithmetic, there exists a prime factorization
$$m = p_1^{n_1} p_2^{n_2} \cdots p_k^{n_k} \tag{1}$$
where $p_1, p_2, \cdots, p_k$ are prime numbers and $n_1, n_2, \ldots, n_k$ are natural numbers. Hence,
$$m^n = (p_1^{n_1} p_2^{n_2} \cdots p_k^{n_k})^n = p_1^{n_1 n} p_2^{n_2 n} \cdots p_k^{n_k n}. \tag{2}$$
Since m^n is divisible by p, there exists an integer d such that $m^n = pd$. Suppose that $q_1^{m_1} q_2^{m_2} \cdots q_i^{m_i}$ is a prime factorization of d. Then
$$m^n = p q_1^{m_1} q_2^{m_2} \cdots q_i^{m_i}. \tag{3}$$
Given (2) and (3), by the uniqueness of the prime factorization of m^n, there exists j (between 1 and k inclusive) such that $p = p_j$. By (1), we know that m is divisible by p_j and, therefore, by p. □

Exercise 26. Prove the converse of Proposition 25.

Use the ideas of Proposition 25 for the following exercise.

Exercise 27. Prove that $\sqrt{3}$ is irrational.

The next exercise is a problem in combinatorics. Counting problems can be notoriously difficult.

Exercise 28. Prove that if at least two people are present at a party, at least two of them know the same number of guests. (You should decide that either everyone or no one counts themselves as someone they know and that if one person knows another, then the latter also knows the former.)

3.4 Proof by Induction

Before we discuss the next example we need some preliminaries. Suppose we want to prove that $n^2 \geq n$ for all natural numbers n. We have to prove infinitely many statements:
$$1^2 \geq 1$$
$$2^2 \geq 2$$
$$3^2 \geq 3$$
$$\cdots.$$

We can denote these statements by

$$P(1)$$
$$P(2)$$
$$P(3)$$
$$\ldots$$

Theorem 29 gives us a way to prove them all.

Theorem 29 (Principle of Induction). *Statements $P(n)$ are true for all $n \in \mathbb{N}$ if*

1. *the first statement, $P(1)$, is true and*
2. *whenever a statement in the sequence, $P(k)$, is true, then the succeeding statement, $P(k+1)$, is also true.*

Remark 30. You may wonder why we used ks instead of ns in the second step of the Principle of Induction. We could have used ns. We wrote ks to remind you that, in this step, the implication if $P(k)$, then $P(k+1)$ has to be proved for every arbitrary but fixed natural number k.

We called the Principle of Induction a theorem. It is close to being an axiom, usually called the Axiom of Infinity. We state this axiom below in a somewhat different (also more readable and less precise) form than you would encounter in an axiomatic approach to set theory.

Axiom 31 (Axiom of Infinity). *There exists a set S of natural numbers that is the smallest set such that*

1. $1 \in S$ *and*
2. *if $k \in S$, then $k + 1 \in S$.*

If we have a sequence of statements

$$P(1), P(2), P(3), P(4), \ldots,$$

then we can define a set

$$S = \{n \mid P(n) \text{ is true}\}.$$

This axiom characterizes the set of natural numbers: $S = \mathbb{N}$. For our purposes, we assume that the Principle of Induction is an axiom. Let us see how this gives us an important proof technique.

The second step in the Principle of Induction can be restated as

$$P(k) \text{ implies } P(k+1) \text{ for all } k \in \mathbb{N}.$$

This step is called the *inductive step* and $P(n)$ is called the *inductive hypothesis* (or *induction hypothesis*). Sometimes it is more advantageous to use the implication $P(k-1)$ implies $P(k)$; then $P(k-1)$ is called the inductive hypothesis. The idea is natural: if $P(1)$ is true and if $P(k)$ implies $P(k+1)$ for all $k \in \mathbb{N}$, then they are all true:

$$P(1) \to P(2) \to P(3) \to \cdots P(k) \to P(k+1) \to \cdots.$$

3.4. Proof by Induction

Remark 32. We like to think of a proof by induction as building a machine: you need an "ON" switch to get it going and then the machine needs to crank out the desired product. The switch is part (1) and the machine mechanism, or crank, is part (2).

We are now ready to apply the Principle of Induction to prove Propositions 33, 34, and 36.

Proposition 33. $n^2 \geq n$ for all $n \in \mathbb{N}$.

Proof. The proof uses the Principle of Induction. For every natural number n, let $P(n)$ denote the statement "$n^2 \geq n$." We first prove $P(1)$. For $n = 1$, we have $1^2 \geq 1$, which is true.

Next we suppose that $P(k)$ is true for some k, that is, $k^2 \geq k$; this is the inductive hypothesis. Next we prove that $P(k+1)$ is true, that is, $(k+1)^2 \geq k+1$. We compute

$$(k+1)^2 = k^2 + 2k + 1$$

and, by the inductive hypothesis,

$$k^2 + 2k + 1 \geq k + 2k + 1 \geq k + 1$$

since $2k \geq 0$. (In fact, $2k \geq 2$ since $k \geq 1$, but we do not need this in the proof. We could say that $k^2 > k$ for $k > 1$. Since $1^2 = 1$, this would slightly improve the statement of Proposition 33.) Therefore, $(k+1)^2 \geq k+1$. □

Do you see that the work in the second paragraph is necessary? Many students who are seeing induction proofs for the first time want to simply assert the truth of $P(k+1)$ by substituting $k+1$ into the inductive hypothesis. Of course, this is wrong.

Proposition 34. $1 + 2 + 3 + \cdots + n = \frac{n(n+1)}{2}$ for all $n \in \mathbb{N}$.

Proof of Proposition 34. The proof is by induction on n. First, we prove the statement for $n = 1$: we have

$$1 = \frac{1(1+1)}{2},$$

which is true.

Next, we suppose that

$$1 + 2 + 3 + \cdots + k = \frac{k(k+1)}{2}.$$

Add $k + 1$ to both sides of our inductive hypothesis and simplify the right-hand side to get

$$1 + 2 + 3 + \cdots + k + (k+1) = \frac{k(k+1)}{2} + (k+1)$$
$$= \frac{k(k+1) + 2(k+1)}{2}$$
$$= \frac{(k+1)(k+2)}{2}.$$

□

Remark 35. The summation notation
$$\sum_{j=m}^{n} f(j)$$
indicates a sum of terms of the form $f(j)$, for some function f, where we substitute sequential integers from m through n in place of j. The statement in Proposition 34 can be written as
$$\sum_{j=1}^{n} j = \frac{n(n+1)}{2}.$$

There is nothing special about starting an induction proof from $n = 1$; it can start with any integer and continue with all integers greater than the original integer. (Even wilder possibilities exist!) To understand this you should realize that it is just a matter of relabeling which integer is "the first."

Proposition 36. $\sum_{j=0}^{n} r^j = \dfrac{1 - r^{n+1}}{1 - r}$ for $n = 0, 1, 2, \ldots$ and $r \neq 1$. (For $r = 0$, use the convention $r^0 = 1$.)

Proof. The proof is by induction. Let r be a real number different from 1. Let $P(n)$, for $n = 0, 1, 2, \ldots$, be the statement
$$\sum_{j=0}^{n} r^j = \frac{1 - r^{n+1}}{1 - r}.$$

We want to prove $P(0)$; that is, we start with $n = 0$.
$$\sum_{j=0}^{0} r^j = r^0 = 1 = \frac{1-r}{1-r} = \frac{1 - r^{0+1}}{1-r}.$$

Next, we want to show, for $k = 0, 1, 2, \ldots$, that $P(k)$ implies $P(k+1)$. Suppose that
$$\sum_{j=0}^{k} r^j = \frac{1 - r^{k+1}}{1-r};$$
this is the inductive hypothesis. We must show that the it implies
$$\sum_{j=0}^{k+1} r^j = \frac{1 - r^{(k+1)+1}}{1-r} = \frac{1 - r^{k+2}}{1-r}.$$

Now
$$\sum_{j=0}^{k+1} r^j = \left(\sum_{j=0}^{k} r^j\right) + r^{k+1} = \frac{1 - r^{k+1}}{1-r} + r^{k+1}$$
by the inductive hypothesis. Moreover,
$$\frac{1 - r^{k+1}}{1-r} + r^{k+1} = \frac{1 - r^{k+1} + (1-r)r^{k+1}}{1-r}$$
$$= \frac{1 - r^{k+1} + r^{k+1} - r^{k+2}}{1-r} = \frac{1 - r^{k+2}}{1-r}.$$

3.4. Proof by Induction

This is what we needed to show and completes the proof. □

Remark 37. The following remarks concern the style of writing proofs by induction.
(1) Since a proof by induction has two steps, two paragraphs make it more readable.
(2) Usually, the first step is much easier than the inductive step.
(3) During a lecture, a mathematician may give a shorthand proof that looks like the following, where "$n = 1$" or "$n = 0$" introduces the first step, "I.H." denotes the inductive hypothesis (it is sometimes left out), and "$k + 1$" introduces the inductive part.

Short proof of Proposition 33. $\boldsymbol{n = 1}$: $1^2 \geq 1$.
  I.H.: $k^2 \geq k$.
  $\boldsymbol{k + 1}$: $(k+1)^2 = k^2 + 2k + 1 \geq k + 2k + 1 \geq k + 1$. □

Short proof of Proposition 34. $\boldsymbol{n = 1}$: $1 = \dfrac{1 \cdot (1+1)}{2}$.
  I.H.: $1 + 2 + 3 + \cdots + k = \dfrac{k(k+1)}{2}$.
  $\boldsymbol{k + 1}$:

$$(1 + 2 + 3 + \cdots + k) + (k+1) = \frac{k(k+1)}{2} + (k+1)$$
$$= \frac{k^2 + 3k + 2}{2} = \frac{(k+1)(k+2)}{2}.$$

□

Short proof of Proposition 36. $\boldsymbol{n = 0}$: $\displaystyle\sum_{j=0}^{0} r^j = r^0 = 1 = \dfrac{1-r}{1-r}$.
  I.H.: $\displaystyle\sum_{j=0}^{k} r^j = \dfrac{1 - r^{k+1}}{1 - r}$.
  $\boldsymbol{k + 1}$:

$$\sum_{j=0}^{k+1} r^j = \left(\sum_{j=0}^{k} r^k\right) + r^{k+1} = \frac{1 - r^{k+1}}{1 - r} + r^{k+1}$$
$$= \frac{1 - r^{k+1} + (1 - r)r^{k+1}}{1 - r} = \frac{1 - r^{k+2}}{1 - r}.$$

□

Can you follow the three proofs above? Do you find the original, more wordy proofs more readable? Written words can help the reader to understand your proofs. We do not support shorthand for its own sake, but it is acceptable as long as there is no loss of understanding such as in a class where the instructor who is writing is there to answer questions!

Remark 38. As we mentioned, it is tempting, when learning how to construct proofs by induction, to simply substitute $k + 1$ for k in the Inductive Hypothesis and then declare victory. That is wrong. Consider the following Absurdum. (For a natural number n, n factorial is the number $n! = n(n-1)(n-2)\cdots(1)$.)

Absurdum 39. $2^{n-1} = n!$ *is true for all* $n \in \mathbb{N}$.

Poof? The "poof" is by mathematical induction. For $n = 1$, we see that
$$2^{1-1} = 2^0 = 1 = 1!.$$
To prove the inductive step, let us suppose that $2^{k-1} = k!$ for some $k \in \mathbb{N}$; this is the inductive hypothesis. Then by substituting $k+1$ in for k in the inductive hypothesis, we get
$$2^{(k+1)-1} = (k+1)!$$
This is what we wanted to show. □

This shows that merely substituting $k+1$ for k does not utilize the inductive hypothesis, $P(k)$, to prove $P(k+1)$.

The following statement and "poof" are wrong.

Absurdum 40. $2^n < n!$ *for all* $n \in \mathbb{N}$.

Poof? The "poof" is by mathematical induction. Suppose that $2^k < k!$ is true for some k; this is our inductive hypothesis. We need to show this implies that $2^{k+1} < (k+1)!$. We see that
$$(k+1)! = (k+1)(k!) > (k+1)(2^k) \geq (1+1)(2^k) = (2)(2^k) = 2^{k+1}.$$
Hence we have shown that $2^{(n+1)} < (n+1)!.$ □

> **Exercise 41.** (a) Explain why the statement of Absurdum 40 is absurd and why its "poof" is wrong.
> (b) Prove that $2^n < n!$ for $n = 4, 5, 6, 7, 8, \ldots$.

Now it is time for you to try a proof by induction.

> **Exercise 42.** Prove that, for all natural numbers n,
> $$\sum_{j=1}^{n} \frac{1}{j(j+1)} = \frac{n}{n+1}.$$

3.5 Proof by Complete Induction

We now state the Principle of Complete Induction, which is an equivalent version of the Principle of Induction.

Theorem 43 (*Principle of Complete Induction*). *Statements $P(n)$ are true for all $n \in \mathbb{N}$ if*

1. $P(1)$ is true and

2. for every $n \in \mathbb{N}$ if $P(k)$ is true for $k = 1, 2, 3, \ldots, n$, then $P(n+1)$ is also true.

3.5. Proof by Complete Induction

The next result is amazing at first glance; we prove it by complete induction. The *Fibonacci sequence* is defined inductively (or recursively) by

$$a_n = \begin{cases} 1, & \text{if } n = 1, 2 \\ a_{n-1} + a_{n-2}, & \text{if } n = 3, 4, 5, \ldots \end{cases}$$

That is, the Fibonacci sequence is

$$\{1, 1, 2, 3, 5, 8, 13, 21, 34, 55, 89, 144, 233, 377, \ldots\}.$$

You probably do not see a formula to determine the thousandth term in this sequence as you would for an arithmetic or geometric sequence. However, there is a formula.

Theorem 44. *The nth term of the Fibonacci sequence is*

$$a_n = \frac{\left(1 + \sqrt{5}\right)^n - \left(1 - \sqrt{5}\right)^n}{2^n \sqrt{5}}.$$

What makes this amazing is the appearance of irrational numbers—square roots of 5—in this sequence of sums of pairs of natural numbers! The number

$$\frac{1 + \sqrt{5}}{2}$$

is called the *golden ratio*. It has interesting properties and comes up in surprising situations, like this one. The number $\frac{1-\sqrt{5}}{2}$ is referred to as the conjugate of the golden ratio.

Before giving the proof, let us consider what must be done. It is simple to check the formula for $a_1 = 1$. In the inductive step using complete induction we want to use the inductive hypothesis

$$a_k = \frac{\left(1 + \sqrt{5}\right)^k - \left(1 - \sqrt{5}\right)^k}{2^k \sqrt{5}},$$

for $k = 1, 2, 3, \ldots, n - 1$, to prove that

$$a_n = \frac{\left(1 + \sqrt{5}\right)^n - \left(1 - \sqrt{5}\right)^n}{2^n \sqrt{5}}.$$

We know that

$$a_n = a_{n-1} + a_{n-2}.$$

By the inductive hypothesis for $k = n - 1$ and $k = n - 2$, we have

$$a_n = \frac{\left(1 + \sqrt{5}\right)^{n-1} - \left(1 - \sqrt{5}\right)^{n-1}}{2^{n-1} \sqrt{5}} + \frac{\left(1 + \sqrt{5}\right)^{n-2} - \left(1 - \sqrt{5}\right)^{n-2}}{2^{n-2} \sqrt{5}}.$$

So we wish to verify that

$$\frac{\left(1 + \sqrt{5}\right)^n - \left(1 - \sqrt{5}\right)^n}{2^n \sqrt{5}} \stackrel{?}{=} \frac{\left(1 + \sqrt{5}\right)^{n-1} - \left(1 - \sqrt{5}\right)^{n-1}}{2^{n-1} \sqrt{5}}$$

$$+ \frac{\left(1 + \sqrt{5}\right)^{n-2} - \left(1 - \sqrt{5}\right)^{n-2}}{2^{n-2} \sqrt{5}}.$$

Using algebra, you might derive the equivalent equations:

$$\left(1+\sqrt{5}\right)^n - \left(1-\sqrt{5}\right)^n \stackrel{?}{=} 2\left[\left(1+\sqrt{5}\right)^{n-1} - \left(1-\sqrt{5}\right)^{n-1}\right]$$
$$+ 4\left[\left(1+\sqrt{5}\right)^{n-2} - \left(1-\sqrt{5}\right)^{n-2}\right]$$

$$\left(1+\sqrt{5}\right)^n - 2\left(1+\sqrt{5}\right)^{n-1} - 4\left(1+\sqrt{5}\right)^{n-2}$$
$$\stackrel{?}{=} \left(1-\sqrt{5}\right)^n - 2\left(1-\sqrt{5}\right)^{n-1} - 4\left(1-\sqrt{5}\right)^{n-2}$$

$$\left(1+\sqrt{5}\right)^{n-2}\left[\left(1+\sqrt{5}\right)^2 - 2\left(1+\sqrt{5}\right) - 4\right]$$
$$\stackrel{?}{=} \left(1-\sqrt{5}\right)^{n-2}\left[\left(1-\sqrt{5}\right)^2 - 2\left(1-\sqrt{5}\right) - 4\right]$$

$$\left(1+\sqrt{5}\right)^{n-2}\left[1 + 2\sqrt{5} + 5 - 2 - 2\sqrt{5} - 4\right]$$
$$\stackrel{?}{=} \left(1-\sqrt{5}\right)^{n-2}\left[1 - 2\sqrt{5} + 5 - 2 + 2\sqrt{5} - 4\right]$$

$$\left(1+\sqrt{5}\right)^{n-2}[0] \stackrel{?}{=} \left(1-\sqrt{5}\right)^{n-2}[0]$$
$$0 \stackrel{?}{=} 0$$

You know that since the last equation is true, all of the previous, equivalent equations are true also. You never see a computation with question marks like this in a textbook and you should be careful using them even as a side computation. It is important that every step is reversible! You can rewrite this computation in a better form, going forwards rather than backwards. We will give a less messy proof below.

Does this complete the proof? No, since our proposed inductive step never settles the case of $n = 2$. This is because we used the fact that

$$a_n = a_{n-1} + a_{n-2},$$

which is valid only for $n \geq 3$. The proof is completed by checking this case.

Let us now rewrite the proof. Before doing so, let us consider the golden ratio and its conjugate again. They are the two roots of the quadratic polynomial $x^2 - x - 1$. Let

$$\varphi = \frac{1+\sqrt{5}}{2} \quad \text{and} \quad \hat{\varphi} = \frac{1-\sqrt{5}}{2},$$

so $\varphi^2 - \varphi - 1 = 0$ and $\hat{\varphi}^2 - \hat{\varphi} - 1 = 0$. That is,

$$\varphi^2 = \varphi + 1 \quad \text{and} \quad \hat{\varphi}^2 = \hat{\varphi} + 1.$$

3.5. Proof by Complete Induction

Proof of Proposition 44. We wish to show that

$$a_n = \frac{\left(1+\sqrt{5}\right)^n - \left(1-\sqrt{5}\right)^n}{2^n \sqrt{5}} = \frac{1}{\sqrt{5}}(\varphi^n - \hat{\varphi}^n).$$

For $n = 1$ and $n = 2$, we compute

$$\frac{1}{\sqrt{5}}(\varphi^1 - \hat{\varphi}^1) = \frac{\left(1+\sqrt{5}\right) - \left(1-\sqrt{5}\right)}{2\sqrt{5}} = 1 = a_1$$

and

$$\frac{1}{\sqrt{5}}(\varphi^2 - \hat{\varphi}^2) = \frac{1}{\sqrt{5}}((\varphi + 1) - (\hat{\varphi} + 1))$$

$$= \frac{1}{\sqrt{5}}(\varphi - \hat{\varphi}) = 1 = a_2.$$

For an integer $n \geq 3$, suppose that the inductive hypothesis

$$a_k = \frac{\left(1+\sqrt{5}\right)^k - \left(1-\sqrt{5}\right)^k}{2^k \sqrt{5}} = \frac{1}{\sqrt{5}}\left(\varphi^k - \hat{\varphi}^k\right)$$

holds for $k = 1, 2, 3, \ldots, n-1$. By the definition of the Fibonacci sequence and the induction hypotheses for $k = n-1$ and $k = n-2$,

$$a_n = a_{n-1} + a_{n-2} = \frac{1}{\sqrt{5}}(\varphi^{n-1} - \hat{\varphi}^{n-1}) + \frac{1}{\sqrt{5}}(\varphi^{n-2} - \hat{\varphi}^{n-2}).$$

To show that $a_n = \frac{1}{\sqrt{5}}(\varphi^n - \hat{\varphi}^n)$, we need to show that

$$\frac{1}{\sqrt{5}}(\varphi^n - \hat{\varphi}^n) = \frac{1}{\sqrt{5}}(\varphi^{n-1} - \hat{\varphi}^{n-1}) + \frac{1}{\sqrt{5}}(\varphi^{n-2} - \hat{\varphi}^{n-2}).$$

We compute that

$$(\varphi^{n-1} - \hat{\varphi}^{n-1}) + (\varphi^{n-2} - \hat{\varphi}^{n-2}) = \varphi^{n-2}(\varphi + 1) - \hat{\varphi}^{n-2}(\hat{\varphi} + 1)$$

$$= \varphi^{n-2}(\varphi^2) - \hat{\varphi}^{n-2}(\hat{\varphi}^2)$$

$$= \varphi^n - \hat{\varphi}^n.$$

Dividing both sides by $\sqrt{5}$ yields the result. □

Let's look at an erroneous attempt at using complete induction.

Absurdum 45. *For all $n \in \mathbb{Z}_+$, $n = 0$.*

Poof? The "poof" is by complete induction. Let $P(n)$ be the predicate $n = 0$. Since the least number in $\mathbb{Z}_+$ is 0, the first step is for $n = 0$. This proves the first step.

For the inductive step, we begin with the induction hypothesis that $P(k)$ is true for $k = 0, 1, 2, \ldots, n$. That is, for $k = 0, 1, 2, \ldots, n$, $k = 0$. We want to use this to show that $n + 1 = 0$. Since, by hypothesis, $1 = 0$ and $n = 0$, then $n + 1 = 0 + 0 = 0$. This completes the poof of the statement. □

There is a problem in the "poof," but where is it? The hypothesis $1 = 0$ is nonsense. However, that alone does not explain how the logic of the proof is faulty. We proved only one starting step,

$$n = 0,$$

but use two hypotheses,

$$1 = 0 \text{ and } n = 0,$$

in proving the inductive step. This brings us closer to the error.

The first application of the inductive step is when $n + 1 = 1$ or, equivalently, when $n = 0$. The inductive hypothesis is $k = 0$ for $k = 0, 1, 2, \ldots, n$, which is just $0 = 0$ when $n = 0$. That is, $1 = 0$ is not part of the inductive hypothesis when $n = 0$.

You can use the Principle of Complete Induction to prove the following, which is the existence part of Theorem 24, the Fundamental Theorem of Arithmetic.

> **Exercise 46.** Prove that every integer n greater than 1 equals a product of one or more prime numbers.

The uniqueness part of the Fundamental Theorem of Arithmetic is not so easy to write down. Think about why it is true and compare your thoughts with the following.

Proof of Theorem 24. In Exercise 46, we proved the existence of such prime factorizations. The Fundamental Theorem of Arithmetic guarantees existence and uniqueness. It remains to prove the uniqueness of prime factorizations.

Assume that there are two different prime factorizations of an integer n with $n \geq 2$:

$$n = p_1 p_2 p_3 \cdots p_a = q_1 q_2 q_3 \cdots q_b;$$

that is, n is factored into a primes on one hand and b primes on the other. Since p_1 is a divisor of $n = q_1 q_2 \cdots q_b$, p_1 must be a divisor of a prime q_j; in fact, $p_1 = q_j$. Rearrange the factorization into q_k's as $q_1' q_2' \cdots q_b'$ with $q_1' = q_j$. Canceling $p_1 = q_1'$ yields

$$p_2 p_3 \cdots p_a = q_2' q_3' \cdots q_b'.$$

Continue to cancel primes in this way until one side of the equation equals 1. Since 1 is neither prime nor the product of primes greater than 1, both sides must equal 1. Therefore, $a = b$. This shows that the factorization is unique up to the order of the factors. The Fundamental Theorem of Arithmetic is proved. $\square$

3.6 Examples and Counterexamples

Thus far we have examined methods of proof. We will now examine a method for disproving statements (or proving negations). The following is a negated statement.

> There do not exist integers a, b, and c such that $a^2 + b^2 = c^2$.

This claim is false, since $a = 3$, $b = 4$, and $c = 5$ satisfies the relation. We say that we have provided a *counterexample*.

3.6. Examples and Counterexamples

Exercise 47. Find a counterexample to the assertion that the absolute value of every real number is positive.

Exercise 48. Find a counterexample to the assertion that every continuous function is differentiable.

As we mentioned in Section 3.1, we consider examples and counterexamples to be extremely important. You should always examine them; often they will help you to find a proof of a statement. You should keep in mind that you can never prove or disprove a statement simply by giving an example unless

1. you examine all possible examples (proof by cases), or

2. the statement is an existence theorem, or

3. the statement is false in at least one case (counterexample).

A counterexample to a statement is also a proof of the negation of the statement. The negation of a statement of the form

$$\text{for all } x, \, P(x)$$

is a one of the form

$$\text{there exists an } x \text{ such that not } P(x),$$

so a counterexample for the statement gives an example that proves a corresponding existence theorem.

For example, consider the existence theorem:

Proposition 49. *There exists a continuous function f that is not differentiable.*

Proof. Your counterexample in Exercise 48 proves it. □

3.7 Supplemental Exercises

Definition Review. There were no definitions in this chapter.

Supplemental Exercise 1. Prove that $2^n + 2 \leq 2^{n+1}$ for all $n \in \mathbb{N}$. Use this result to prove that $2n + 1 \leq 2^n$ for $n = 3, 4, 5, \ldots$.

Supplemental Exercise 2. Prove that the sum of an even integer and an odd integer is odd.

Supplemental Exercise 3. Prove that the cube of an odd integer is odd.

Supplemental Exercise 4. Explain why the statement of the following Absurdum is absurd and how its "poof" is wrong.
  Absurdum. If $x > 0$, then for all $y > 0$, $xy = 1$.
  Poof? We proceed by contradiction. Assume that for all $y > 0$, $xy \neq 1$. Since $0y = 0 \neq 1$, $x = 0$ satisfies this condition. This contradicts the fact that $x > 0$. □

Supplemental Exercise 5. Prove that the sum of two rational numbers is a rational number.

Supplemental Exercise 6. Prove that the difference of two rational numbers is a rational number.

Supplemental Exercise 7. Prove that the product of two rational numbers is a rational number.

Supplemental Exercise 8. Prove that the quotient (when it is defined) of two rational numbers is a rational number. State the condition that you use to guarantee the existence of the quotient.

Supplemental Exercise 9. It can be shown that the product of positive reals is positive and the product of negative reals is positive. Use these facts to prove that, for all real numbers a, if $a \neq 0$, then $a^2 > 0$.

Supplemental Exercise 10. Prove that $\sqrt{5}$ is irrational.

Supplemental Exercise 11. Prove that the positive square root of every prime number is irrational.

Supplemental Exercise 12. At a party, 25 guests mingle and shake hands with some fellow guests. Prove that at least one guest must have shaken hands with an even number of guests.

Supplemental Exercise 13. Let a and b be real numbers. Prove that $a^2 + b^2 = 0$ if and only if $a = b = 0$.

Supplemental Exercise 14. Let x be a real number. Prove that if $x^2 < x$, then $x < 1$.

Supplemental Exercise 15. Explain why the statement of the following Absurdum is absurd and how its "poof" is wrong.
  Absurdum. For $n = 1, 2, 3, \ldots$, every bag containing n solid-colored balls contains balls of only one color.

3.7. Supplemental Exercises

Poof? **n = 1:** A bag with one ball clearly has balls of only one color.

n + 1: Take a bag with $n + 1$ balls. Remove one ball. By hypothesis, the bag now has balls of only one color. Replace the first ball and remove a different ball. Again, the bag now has balls of only one color. Since the other balls that never left the bag did not change color, all the balls must be the same color. □

Supplemental Exercise 16. Prove that $1 + 3 + 5 + \cdots + (2n - 1) = n^2$ for all $n \in \mathbb{N}$.

Supplemental Exercise 17. Prove that, for all natural numbers n,

$$\sum_{k=1}^{n} k^2 = \frac{n(n+1)(2n+1)}{6}.$$

Supplemental Exercise 18. Give a proof by induction for Exercise 28.

Supplemental Exercise 19. Let the sequence $\{a_n\}$ be defined by: $a_1 = 3$ and

$$a_{n+1} = \frac{3a_n + 1}{4}$$

for $n = 2, 3, \ldots$. Prove that for every natural number n, $a_{n+1} < a_n$.

Supplemental Exercise 20. Let the sequence $\{a_n\}$ be defined by: $a_1 = 3$ and

$$a_{n+1} = \frac{3a_n + 1}{4}$$

for $n = 2, 3, \ldots$. Prove that for every natural number n, $a_n > 1$.

Supplemental Exercise 21. Prove that, for every natural number n, $6^n - 1$ is a multiple of 5.

Supplemental Exercise 22. Suppose that x is a real number in the interval $[0, \pi/2]$. Prove that

$$\sin(nx) \leq n \sin(x)$$

for every natural number n.

Supplemental Exercise 23. Suppose that D is an operation taking functions on the reals to functions on the reals. Assume that $D(x) = 1$ and that

$$D(f(x)g(x)) = D(f(x))g(x) + f(x)D(g(x))$$

for all functions $f(x)$ and $g(x)$. Prove that $D(x^n) = nx^{n-1}$ for every natural number n.

Supplemental Exercise 24. Prove that for every natural number n

$$\sum_{i=0}^{n}(2i + 1) = (n + 1)^2.$$

Supplemental Exercise 25. Prove that for every natural number n

$$\sum_{j=0}^{n} \frac{1}{j(j+1)} = \frac{n}{n+1}.$$

Supplemental Exercise 26. Suppose that there are three people, two white hats, and three red hats. Each person wears one of the hats, but does not know its color. The first of the three can see the hats worn by the other two. When asked if he knows the color of his hat, he answers "No." The second can see only the hat of the third person. When asked if knows the color of his hat, he also answers "No." The third cannot see any hats. When asked if knows the color of his hat, he answers "Yes!" What color is the hat of the third person? Prove it!

Supplemental Exercise 27. In this exercise, there are two variables, p and n. Induction is on one variable that varies through integer values. Prove that for every real number $p \geq -1$ and any natural number n,
$$(1+p)^n \geq 1 + np.$$

Supplemental Exercise 28. In Supplemental Exercise 27, what happens when $p < -1$? Is the statement still true? If not, determine where the proof breaks down.

Supplemental Exercise 29. Explain why the statement of the following Absurdum is absurd and how its "poof" is wrong.
Absurdum. For all $n \in \mathbb{Z}_+$, $10^n = 1$.
Poof? We use complete induction.
n = 0: $10^0 = 1$.
n + 1: Since $10^n = 10^{n-1} = 1$,
$$10^{n+1} = 10^{2n-(n-1)} = \frac{10^{2n}}{10^{n-1}} = \frac{(10^n)^2}{10^{n-1}} = \frac{1^2}{1} = 1.$$

Supplemental Exercise 30. Let a_k be the terms of the Fibonacci sequence. (That is, $a_1 = a_2 = 1$ and $a_k = a_{k-2} + a_{k-1}$ for all $k = 3, 4, 5, \ldots$.) Use complete induction to prove that $a_k < 0$ for all natural numbers k.

Supplemental Exercise 31. Let a_k be the terms of the Fibonacci sequence. Use complete induction to prove that $a_{k+1} \geq a_k$ for all natural numbers k.

Supplemental Exercise 32. Let a_k be the terms of the Fibonacci sequence. Use complete induction to prove that $a_{k+1} \leq 2a_k$ for all natural numbers k.

Supplemental Exercise 33. Prove that $a^2 + ab + b^2 > 0$ for all nonzero real numbers a and b.

Supplemental Exercise 34. Prove that, for real numbers a and b, if $a + b \geq 1$, then $a \geq \frac{1}{2}$ or $b \geq \frac{1}{2}$.

Supplemental Exercise 35. Prove that there is only one right triangle whose sides have lengths that are consecutive positive integers.

Supplemental Exercise 36. Prove that $|x^{2011} - x - 1| > -x^2 - 1$ for all $x \in \mathbb{R}$.

Supplemental Exercise 37. Find an irrational number $x < 5$ such that
$$(4-x)(y^{2n} + 1) > 0,$$
for all $y \in \mathbb{R}$ and $n \in \mathbb{Z}_+$.

3.7. Supplemental Exercises

Supplemental Exercise 38. Prove that for all positive integers, n,
$$\frac{1}{1^2} + \frac{2}{2^2} + \cdots + \frac{n}{n^2} \leq 2 - \frac{1}{n}.$$

Supplemental Exercise 39. Prove that for all positive integers n with $n \geq 3$,
$$n^{n+1} > (n+1)^n.$$

Supplemental Exercise 40. Prove that if b and c are odd integers, then $x^2 + bx + c = 0$ does not have integer solutions.

Supplemental Exercise 41. Let $f : \mathbb{Z}_+ \to \mathbb{Z}_+$ be a function such that
$$f(n+1) = f(n) + f(1),$$
for all $n \in \mathbb{Z}_+$. Prove that
(a) $f(0) = 0$,
(b) $f(n) = nf(1)$, and
(c) $f(m+n) = f(m) + f(n)$ for all $m, n \in \mathbb{Z}_+$.

Supplemental Exercise 42. Determine whether each of the following statements is true or false. Prove your answers.
1. for all $k \in \mathbb{R}$, there exists $n \in \mathbb{N}$ such that $|n^2 + n - 2| < k$
2. for all $k \in \mathbb{R}$, for all $n \in \mathbb{N}$ such that $|n^2 + n - 2| < k$
3. there exists $k \in \mathbb{R}$ such that for all $n \in \mathbb{N}$ such that $|n^2 + n - 2| < k$
4. there exists $k \in \mathbb{R}$ such that there exists $n \in \mathbb{N}$ such that $|n^2 + n - 2| < k$

Supplemental Exercise 43. Let $P(a)$ be the predicate "$a + \frac{1}{a} < -7$" and $Q(a)$ be the predicate "$\sum_{i=1}^{2011} a^i = 0$." Determine the truth value of the implication $P(a) \Rightarrow Q(a)$ for all $a \in \mathbb{R}$.

Supplemental Exercise 44. Project. Investigate the role and function of proof in mathematics.

How to THINK about mathematics: A Summary

Before trying to prove a statement:
- Construct examples (under same and modified hypotheses).
- Construct counterexamples (same and modified hypotheses).
- Construct pictures (if applicable).

Consider different proof techniques:
- Direct Proof (including Brute Force).
- Proof by Contraposition (or proof of the contrapositive).
- Proof by Contradiction.
- Proof by Induction (including Complete Induction).

Consider proof by cases and breaking a proof into smaller pieces.

How to COMMUNICATE mathematics: A Summary

Discuss mathematics with people—all parties should be honest and active.
Organize your ideas (rewrite your proofs several times, if necessary).
Remember to use enough words (a reader of your written proofs cannot ask you questions).
Use notation only as a shorthand or to clarify your proofs.

How to DO mathematics: A Summary

Read with pencil or pen in hand and use it.
Write down your ideas.
Talk out your ideas (have a second party present!).
Think (see above).
Communicate (see above).

SET THEORY

4
Basic Set Operations

4.1 Introduction

Perhaps our ability to distinguish and to classify objects constitutes the essential part of our intellectual armor as human beings. These objects may be tangible—inanimate objects, distant galaxies, living beings, our beloved persons—or they may be ideas, feelings, or figments of our imagination. Often we view a collection of objects as a unity. When Aristotle claimed that man is a rational animal, he meant to say that all human beings are rational. In fact, of all the animals, he isolated those, and only those, who shared a common characteristic. Categorization is not always an easy process. Often, it stumbles over insurmountable ambiguities, especially due to the fact that universal agreement on the meaning of terms is not always achieved.

If we restrict reality to the mathematical realm, it turns out that we can come up with an elegant and powerful theory. The key concept is that of a set. To cite Cantor's so-called naive definition, a set is any collection of definite, distinguishable objects of our intuition or of our intellect to be conceived as a whole.

A set may be specified in different ways. One way is to list the elements of a set explicitly as in

$$T = \{ \text{earth, wind, fire, water} \}.$$

The set T has exactly four elements. Similarly, we can list enough elements so that a pattern emerges that describes all of the elements of the set such as

$$E = \{ 2, 4, 6, \ldots \}.$$

The set of all even natural numbers, E, has infinitely many elements; the three dots indicate that the list continues indefinitely in the same pattern.

We may specify E in different ways as follows:

$$\begin{aligned} E &= \{ n \in \mathbb{N} \mid 2 \text{ divides } n \} \\ &= \{ 2n \mid n \in \mathbb{N} \}. \end{aligned}$$

This kind of notation is called set-builder notation. It is common to read this notation by substituting "such that" for the vertical bar. Some authors prefer to use a colon in place of the vertical bar. The set $\mathbb{N}$ appears on different sides of the vertical bar. We must specify a set for every variable.

4.2 Subsets

To study set theory, as with any new mathematics, we start with some undefined terms and some definitions. Since a sentence introducing defined terms is called a definition, we can call a sentence introducing undefined terms an undefinition. Or perhaps we should call it a primativum since primitive terms are introduced?

Undefinition 1. The terms *set*, *element*, and *is an element of* (represented by the symbol $\in$) are undefined.

For example, the notation $x \in S$ means that x is an element of a set S. We will be sloppy with this notation when we write something like $x, y \in S$ to mean $x \in S$ and $y \in S$.

Remark 2. Naively, a set is a collection of objects. When listing the elements of a set, it is traditional that no element is repeated. For example, $\{1, 2\}$ is a set, but $\{1, 1\}$ is not a set, unless we think that, as sets, $\{1, 1\} = \{1\}$.

When elements are allowed to repeat, we use the term *multiset*. So, $\{1, 1\}$ is a multiset. Every set can be thought of as a multiset, including the empty set, which we define below.

Sets are frequently called *collections*, but the naive notion that every collection is a set is examined further in Section 4.7.

To make set theory interesting, we need an axiom that tells us that some set actually exists. The existence of the following set is assumed axiomatically.

Definition 3. A set is *empty* if and only if the set contains no elements. The empty set is denoted in two ways, $\emptyset$ and $\{\}$.

You should be careful about notation! The sets $\{\{\}\}$ and $\{\emptyset\}$ are not empty. Both have exactly one element, namely the empty set. On the other hand, an empty set has zero elements!

You may have noticed that we wrote "the" empty set instead of "an" empty set in Definition 3. This will be justified in Proposition 15; there is only one empty set.

Definition 4. A set S is a *subset* of a set X if and only if every element of S is also an element of X. We write either $S \subset X$ or $X \supset S$.

Remark 5. By Definition 4, $S \subset X$ is equivalent to $s \in S$ implies $s \in X$. In general, to prove that $S \subset X$, choose an arbitrary element s of S and show that s must be in X also. We call this *chasing elements*.

Remark 6. You should be careful with the notation used in this text. Many authors, especially of more elementary texts, use the notation $S \subseteq X$ to describe subsets. They use $S \subset X$ to mean that S is a *proper* subset of X, that is, S is a subset of X, but is not equal to X. You may take the concept of set equality for granted. However, we will define it in Definition 11 later in this section. We will use the notation $S \subsetneq X$ for proper subsets. Because it is more common to discuss subsets than proper subsets, we use the simpler notation for the more common usage. In more advanced mathematical texts, it is common to use $\subset$ for all subsets.

4.2. Subsets

There is a moral to this story: whenever you pick up a new mathematics text, you should be sure that you understand the notation and the terminology since usage may differ among different authors.

In constructing proofs, it is often helpful to draw pictures, as we noted in Section 3.1 of Chapter 3. For proofs about sets, Venn diagrams, such as in Figure 4.1, may help you to construct proofs.

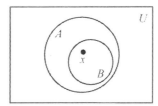

Figure 4.1. Example of a Venn diagram

There is no geometry in Figure 4.1; that the shapes are circles is irrelevant. The sets A and B are subsets of U. Also, B is a subset of A. Sometimes it is useful to indicate elements explicitly. The element x is an element of B (and of A and of U). When x is not an element of A, we write $x \notin A$. Remember Venn diagrams when you construct proofs in this chapter.

Absurdum 7. *Let X be a set. The empty set is not a subset of X.*

Poof!? For the empty set to be a subset of X, every element of the empty set must belong to X. Since the empty set has no elements, this condition is not satisfied. Therefore, the empty set is not a subset of X. □

Exercise 8. Find the fallacy in the "poof" of Absurdum 7 and prove that $\emptyset \subset X$.

Keep in mind that the notation $A \subset B$ does not imply that $A \neq B$ (we have not even defined the equality—or inequality—of sets yet). Keeping this in mind, prove the following:

Exercise 9. Let X be a set. Prove that $X \subset X$.

The following important property is called the transitive property of subsets.

Exercise 10. Let A, B, and C be sets. Prove that if $A \subset B$ and $B \subset C$, then $A \subset C$.

If you ask yourself what it means for two sets A and B to be equal, you might answer that A and B have exactly the same elements. We make this precise in the following definition.

Definition 11. Two sets A and B are *equal* if and only

$$x \in A \text{ if and only if } x \in B.$$

When A equals B we write $A = B$.

Proving that two sets are equal can be easy, as in the following example.

Example 12. Let $A = \{2, 4, 6\}$ and $B = \{2n \mid n \in \mathbb{Z}_3\}$, where $\mathbb{Z}_3 = \{1, 2, 3\}$. We see that the elements of B are $2(1) = 2$, $2(2) = 4$, and $2(3) = 6$. These are exactly the elements of A. Hence, $A = B$.

However, proving that two sets are equal is usually much more difficult than in Example 12! The following is a useful way of thinking about equality of sets.

Exercise 13. Let A and B be sets. Prove that $A = B$ if and only if $A \subset B$ and $A \supset B$.

That is, to prove that two sets A and B are equal, we must show that $a \in A$ implies $a \in B$, and $b \in B$ implies $b \in A$.

Remark 14. We can negate subset and set equality. For sets A and B, we write $A \not\subset B$ and $A \neq B$. By Exercise 13, $A \neq B$ if and only if $A \not\subset B$ or $A \not\supset B$.

When A is a subset of B, but $A \neq B$, we write $A \subsetneq B$ and we say that A is a *proper subset* of B. See Remark 6.

You may have noticed we wrote "the" empty set (and not "an" empty set) in and after Definition 3. The following proposition shows that there is a unique empty set.

Proposition 15. *If A and B are empty sets, then $A = B$.*

Proof. By Exercise 13, to show that $A = B$, it suffices to show that $A \subset B$ and $B \subset A$. By Exercise 8, since A is empty, $A \subset B$. Again by Exercise 8, since B is empty, $B \subset A$. □

In the following exercise, we examine the important properties of equality, called the reflexive, symmetric, and transitive properties. They characterize equality in a sense that we will explore in detail in Chapter 6.

Exercise 16. Let A, B, and C be sets. Prove that
(a) $A = A$,
(b) if $A = B$, then $B = A$, and
(c) if $A = B$ and $B = C$, then $A = C$.

4.3 Intersections and Unions

Next we consider two basic set operations, intersection and union. We will do this first for two sets and then for arbitrarily many sets. The intersection of two sets A and B is the set consisting of all elements common to both of the sets A and B.

Definition 17. Let A and B be sets. The *intersection* of A and B is the set

$$A \cap B = \{x \mid x \in A \text{ and } x \in B\}.$$

Remark 18. Definition 17 implies that $x \in A \cap B$ if and only if $x \in A$ and $x \in B$. This equivalence is useful when chasing elements: if you know that some x is an element of $A \cap B$, then x is an element of *both* A and B and you continue chasing from there.

4.3. Intersections and Unions

Notice the similarity in notation for intersection ($\cap$) and the logical connective "and" ($\wedge$).

Consider how you see intersections in Venn diagrams, such as in Figure 4.2. The left

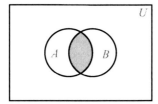 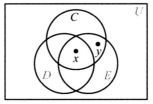

Figure 4.2. Intersections in Venn diagrams

figure shows two sets, A and B with the intersection $A \cap B$ shaded. The right figure shows three sets intersecting each other in the most general way possible. So, $x, y \in C \cap E$, $x \in (C \cap E) \cap D$ but $y \notin (C \cap E) \cap D$.

Let us consider two intersection properties.

Exercise 19. Let A be a set. Prove that $A \cap \emptyset = \emptyset$.

Exercise 20. Let A be a set. Prove that $A \cap A = A$.

Definition 21. Two sets A and B are *disjoint* if and only if $A \cap B = \emptyset$.

By Exercise 19, every set is disjoint from the empty set. Also, by Proposition 15 and Exercise 20, only the empty set is disjoint from itself.

Next, we consider unions. The union of A and B is the set consisting of all elements of A and B.

Definition 22. Let A and B be sets. The *union* of A and B is the set

$$A \cup B = \{x \mid x \in A \text{ or } x \in B\}.$$

Remark 23. Definition 22 implies that $x \in A \cup B$ if and only if $x \in A$ or $x \in B$. This equivalence is useful when chasing elements: if you know that some x is an element of $A \cup B$, then x is an element of A or B. (Remember that "or" could mean both!) You continue chasing from there.

Notice the similarity in notation for union ($\cup$) and the logical connective "or" ($\vee$).

An element may appear only once in a set. So $\{1, 2\} \cup \{2, 3\} = \{1, 2, 3\}$; in fact, $\{1, 2, 2, 3\}$ is not a set.

The Venn diagrams in Figure 4.3 show the unions of two and three sets, respectively. Here are some exercises.

Exercise 24. Let A be a set. Prove that $A \cup \emptyset = A$.

Exercise 25. Let A be a set. Prove that $A \cup A = A$.

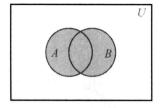 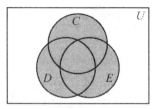

Figure 4.3. Unions in Venn diagrams

The following exercise relates intersections and unions. Note that $X \subset Y \subset Z$ if and only if $X \subset Y$ and $Y \subset Z$.

Exercise 26. Let A and B be sets. Prove that $A \cap B \subset A \subset A \cup B$. Prove that this implies that $A \cap B \subset A \cup B$.

Remember that $A \cap B \subset A \cup B$ does not necessarily mean that $A \cap B$ is a proper subset of $A \cup B$; this should remind you of the non-exclusiveness of the logical "or" operation.

Exercises 19 and 20 are generalized by Proposition 27; Exercises 24 and 25 are generalized by Supplemental Exercise 13.

Proposition 27. *Let A and B be sets. $A \cap B = A$ if and only if $A \subset B$.*

Proof. Suppose that $A \cap B \supset A$. By Exercise 26, $A \cap B \subset B$. Therefore, by Exercise 10, $A \subset B$.

Conversely, suppose that $A \subset B$. Let $x \in A$. Since $A \subset B$, $x \in B$ also. Therefore, $x \in A \cap B$ and $A \subset A \cap B$. This proves that $A \cap B \supset A$ if and only if $A \subset B$.

Let us consider the second part. Suppose that $A \cap B = A$. Thus, $A \cap B \supset A$, which implies that $A \subset B$.

Conversely, suppose that $A \subset B$. Hence, $A \cap B \supset A$. By Exercise 26, $A \cap B \subset A$. Therefore, $A \cap B = A$. □

We next consider the commutative, associative, and distributive properties of intersections and unions. First, we prove the commutative property for intersections.

Proposition 28. *For sets A and B, $A \cap B = B \cap A$.*

Proof. Let $x \in A \cap B$. This is equivalent to $x \in A$ and $x \in B$, which is equivalent to $x \in B$ and $x \in A$. This is equivalent to $x \in B \cap A$. Therefore, $A \cap B = B \cap A$. □

We have commutative properties for intersections and unions (see Proposition 28 and Supplemental Exercise 15):

$$A \cap B = B \cap A \quad \text{and} \quad A \cup B = B \cup A.$$

We have associative properties for intersections and unions (see Supplemental Exercise 16):

$$(A \cap B) \cap C = A \cap (B \cap C)$$

and

$$(A \cup B) \cup C = A \cup (B \cup C).$$

4.4. Intersections and Unions of Arbitrary Collections

We have distributive properties for intersections and unions (see Supplemental Exercise 17):

$$A \cap (B \cup C) = (A \cap B) \cup (A \cap C)$$

and

$$A \cup (B \cap C) = (A \cup B) \cap (A \cup C).$$

By the commutative and distributive properties above, we have the following distributive properties:

$$(A \cup B) \cap C = (A \cap C) \cup (B \cap C)$$

and

$$(A \cap B) \cup C = (A \cup C) \cap (B \cup C).$$

Some people use terms like "left" or "right" distributive properties; we do not use them since we can never remember which is which!

4.4 Intersections and Unions of Arbitrary Collections of Sets

Let us now return to the definitions of intersection and union, this time for a collection of sets. The intersection of a collection of sets is the set consisting of the common elements of all of the sets in the collection.

Definition 29. Let $\mathcal{A}$ be a collection of sets. The *intersection* of the sets in $\mathcal{A}$ is the set

$$\{a \mid a \in A \text{ for all } A \in \mathcal{A}\}.$$

Some authors say the *intersection of* $\mathcal{A}$. The notation is usually

$$\bigcap_{A \in \mathcal{A}} A \quad \text{or} \quad \bigcap \mathcal{A}.$$

Definition 21 defined a disjoint pair of sets. Before generalizing to collections of sets, consider Figure 4.4. It shows two collections, $\mathcal{A}$ and $\mathcal{B}$, of subsets of $\mathbb{R}^2$ that exhibit some kind of disjointness property; in both $\mathcal{A}$ and $\mathcal{B}$, the circular and triangular regions are disjoint as in Definition 21.

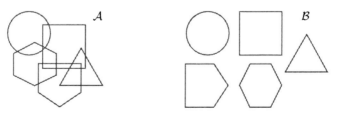

Figure 4.4. Disjointness properties of collections of sets

Definition 30. Let $\mathcal{A}$ be a collection of sets. The sets in $\mathcal{A}$ are *pairwise disjoint* if and only if A_1 and A_2 are disjoint for all $A_1, A_2 \in \mathcal{A}$ such that $A_1 \neq A_2$.

The sets in $\mathcal{A}$ are pairwise disjoint if, for all $A_1, A_2 \in \mathcal{A}$, $A_1 \neq A_2$ implies that
$$A_1 \cap A_2 = \emptyset.$$

Compare this with the following definition.

Definition 31. Let $\mathcal{A}$ be a collection of sets. The sets in $\mathcal{A}$ are *disjoint* if and only if
$$\bigcap \mathcal{A} = \emptyset.$$

This definition of a disjoint collection is not universally used since pairwise disjointness is usually desired. Hence, some authors use disjoint to mean pairwise disjoint. In Figure 4.4, which collection(s) is (are) pairwise disjoint and which collection(s) is (are) disjoint?

Often a collection of sets is indexed by a parameter. For example, the collection of intervals
$$\left\{ (0,1), \left(0, \frac{1}{2}\right), \left(0, \frac{1}{3}\right), \ldots \right\} = \left\{ \left(0, \frac{1}{n}\right) \,\middle|\, n \in \mathbb{N} \right\}$$
is indexed by $\mathbb{N}$. For operations such as intersections of indexed sets the following notations are equivalent:
$$\bigcap_{n=1}^{\infty} \left(0, \frac{1}{n}\right) = \bigcap_{n \in \mathbb{N}} \left(0, \frac{1}{n}\right)$$
$$= (0,1) \cap \left(0, \frac{1}{2}\right) \cap \left(0, \frac{1}{3}\right) \cap \ldots.$$

Example 32. Figure 4.5 shows the nested intervals $(0,1)$, $(0, \frac{1}{2})$, $(0, \frac{1}{3})$, ... Then

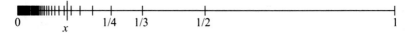

Figure 4.5. Nested intervals with empty intersection

$\bigcap_{n=1}^{\infty} \left(0, \frac{1}{n}\right) = \emptyset$ since $1/n$ can be made arbitrarily small by choosing a sufficiently large n. To be more precise, start by assuming that there exists
$$x \in \bigcap_{n=1}^{\infty} \left(0, \frac{1}{n}\right).$$

Hence, $x > 0$ and, for all $n \in \mathbb{N}$, $x < 1/n$. But this is false since there are natural numbers greater than $1/x$. This obvious statement is called the Archimedean Principle; we will study it more carefully in Section 8.5 of Chapter 8.

Exercise 33. Determine:
(a) $\bigcap_{n=1}^{\infty} [0, \frac{1}{n})$
(b) $\bigcap_{n=2}^{\infty} [\frac{1}{n}, 1)$.

4.4. Intersections and Unions of Arbitrary Collections

Proposition 34. *Let $\mathcal{A}$ be a collection of sets containing at least two sets. If $\mathcal{A}$ is a pairwise disjoint collection, then $\mathcal{A}$ is a disjoint collection.*

Proof. Suppose that $\mathcal{A}$ is a pairwise disjoint collection of sets. Choose distinct $A_1, A_2 \in \mathcal{A}$. Hence, $A_1 \cap A_2 = \emptyset$. Therefore,

$$\bigcap \mathcal{A} = (A_1 \cap A_2) \cap \bigcap \mathcal{A} = \emptyset \cap \bigcap \mathcal{A} = \emptyset.$$

Thus, $\mathcal{A}$ is a disjoint collection. $\square$

Make sure that you understand why

$$\bigcap \mathcal{A} = (A_1 \cap A_2) \cap \bigcap \mathcal{A}$$

in the proof.

The definition of the union of a collection of sets is similar to that of the intersection. The union of the sets in a collection is the set consisting of all of the elements of all of the sets in the collection.

Definition 35. Let $\mathcal{A}$ be a collection of sets. The *union* of the sets in $\mathcal{A}$ is the set

$$\{a \mid \text{there exists } A \in \mathcal{A} \text{ such that } a \in A\}.$$

In analogy with intersections, some authors say the *union of $\mathcal{A}$*. The notation is usually

$$\bigcup_{A \in \mathcal{A}} A \quad \text{or} \quad \bigcup \mathcal{A}.$$

We can rephrase this definition to say that something is an element of the union of $\mathcal{A}$ if and only if there is a set in $\mathcal{A}$ that contains it.

Exercise 36. Determine the following:

(a) $\bigcup_{n=1}^{\infty} (0, \frac{1}{n})$;

(b) $\bigcup_{n=2}^{\infty} [\frac{1}{n}, 1)$.

The distributive properties can be generalized for collections of sets (see Supplemental Exercise 18). What happened to the associative and commutative properties? You can think of these properties as coming for free in the definitions of intersection and union for collections.

Remark 37. Consider the definitions of unions and intersections: for two sets and for collections. "For all" should remind you of "and"; "there exists" should remind you of "or." For example, you might say "for all x and y" and "there exists x or y."

Remark 38. Let us consider the definitions of intersection and union for small collections $\mathcal{A}$. For example, suppose $\mathcal{A} = \{A, B\}$. Then

$$\cap \mathcal{A} = A \cap B \quad \text{and} \quad \cup \mathcal{A} = A \cup B.$$

If $\mathcal{A} = \{A\}$, then
$$\cap \mathcal{A} = A \quad \text{and} \quad \cup \mathcal{A} = A.$$

If $\mathcal{A} = \emptyset$, then, since there does not exist a set in the empty collection, no element can be contained in a set in the empty collection $\mathcal{A}$. Therefore,
$$\cup \mathcal{A} = \emptyset.$$

What about $\cap \mathcal{A}$? Which elements are contained in every set of the empty collection $\mathcal{A}$? Every element satisfies this condition vacuously! That is, the statement if $A \in \mathcal{A}$, then $x \in A$ is true for all elements x. So what is this set of all possible elements? If we assume that every set we are considering is a subset of some *universal set* U, then
$$\cap \mathcal{A} = U.$$

The necessity of universal sets comes up often, e.g., Definition 44 of the complement of a set. It is useful for avoiding paradoxes such as those in Section 4.7. You may have wondered where the elements in Definitions 29 and 35 came from.

4.5 Differences and Complements

We define set difference and use this to define the complement of a set.

Definition 39. Let A and B be sets. The *set difference* $A - B$ is the set of elements of A that are not elements of B.

The notation $A \setminus B$ is commonly used for set difference. It does not look good in handwritten usage. In fact, it can be difficult to distinguish $A \setminus B$, $A \mid B$, and A/B; hopefully the context will make the meaning clear even if the slant is not clear.

The difference of sets can be expressed as
$$A - B = \{x \mid x \in A \text{ and } x \notin B\}.$$

We prove a proposition that characterizes set difference. Let A and B be sets. A subset X of A is the *largest subset of A that is disjoint from B* if and only if, for every set C, if $C \subset A$ and $C \cap B = \emptyset$, then $C \subset X$. The largest subset of A disjoint from the empty set is the largest subset of A, that is A itself.

Exercise 40. Prove that the largest subset of a set A disjoint from $\emptyset$ is A.

We are ready to prove the following.

Proposition 41. *For sets A and B, $A - B$ is the largest subset of A that is disjoint from B.*

Proof. By the definition of set difference, $A - B$ is a subset of A.

Let $C \subset A$ such that $C \cap B = \emptyset$. We want to show that $C \subset A - B$. Choose $x \in C$. Since $C \subset A$, $x \in A$. Since $C \cap B = \emptyset$, $x \notin B$. Therefore, $x \in A - B$ and, hence, C is a subset of $A - B$. $\square$

Set difference is not a commutative operation.

4.6. Power Sets

Exercise 42. (a) Give a counterexample to $A - B = B - A$.
(b) Give an example where $A - B = B - A$.

Set difference is not an associative operation either.

Exercise 43. Give a counterexample to $A - (B - C) = (A - B) - C$.

What is the complement of a set? To define it, we rely on the existence of a universal set, of which all sets are subsets.

Definition 44. Suppose that U is the universal set. The *complement* of A is $U - A$.

The complement is denoted differently by many authors; A', A^c, $\complement A$, and $U - A$ are all used to denote the complement. We will use $U - A$, or A' when U is clear from the context.

4.6 Power Sets

You may not be familiar with the power set of a set. It is an elementary construction that is useful in more advanced topics in mathematics. We will use it again when we discuss different types of infinities in Chapter 7.

Definition 45. Let X be a set. The *power set* of X, denoted $\mathcal{P}(X)$, is the set of all subsets of X.

The simplest power set is the power set of the empty set.

Exercise 46. What is $\mathcal{P}(\emptyset)$? Prove your result.

For more examples,
$$\mathcal{P}(\{1\}) = \{\emptyset, \{1\}\}$$
and
$$\mathcal{P}(\{1,2\}) = \{\emptyset, \{1\}, \{2\}, \{1,2\}\}.$$

Because $\emptyset$ and X are always subsets of X, the power set of X always contains them.

For $n \in \mathbb{N}$, let $\mathbb{Z}_n = \{1, 2, 3, \ldots, n\}$ and $\mathbb{Z}_0 = \emptyset$. We know that
$$\mathcal{P}(\mathbb{Z}_0) = \mathcal{P}(\emptyset) = \{\emptyset\}$$
has one element,
$$\mathcal{P}(\mathbb{Z}_1) = \mathcal{P}(\{1\}) = \{\emptyset, \{1\}\}$$
has two elements, and
$$\mathcal{P}(\mathbb{Z}_2) = \mathcal{P}(\{1,2\}) = \{\emptyset, \{1\}, \{2\}, \{1,2\}\}$$
has four elements. This can be generalized.

Exercise 47. How many elements does $\mathcal{P}(\mathbb{Z}_n)$ have? Prove your result.

Exercise 47 can be proved by induction and by a direct counting argument. Try to give both proofs.

Proposition 48. *If $S \subset X$, then $\mathcal{P}(S) \subset \mathcal{P}(X)$.*

Proof. Suppose that $S \subset X$. Choose an element of $\mathcal{P}(S)$. By the definition of power set, it is a subset A of S. Since $A \subset S$ and $S \subset X$, $A \subset X$. By the definition of power set, $A \in \mathcal{P}(X)$. Hence, $\mathcal{P}(S) \subset \mathcal{P}(X)$. □

It follows from Proposition 48 that if $S = X$, then $\mathcal{P}(S) = \mathcal{P}(X)$.

Absurdum 49. *If A and B are sets, then $\mathcal{P}(A) \cup \mathcal{P}(B) = \mathcal{P}(A \cup B)$.*

Poof!? The following are equivalent:

$$X \in \mathcal{P}(A \cup B)$$
$$X \subset A \cup B$$
$$X \subset A \text{ or } X \subset B$$
$$X \in \mathcal{P}(A) \text{ or } X \in \mathcal{P}(B)$$
$$X \in \mathcal{P}(A) \cup \mathcal{P}(B)$$

□

Exercise 50. Explain why the statement above is absurd and why its proof is wrong.

4.7 Russell's Paradox

We have mentioned "collections" of sets in this chapter. We should be careful. If we were to develop the theory of sets axiomatically, we would need to define this term. Or we might want to replace "collection" with "set." The use of sets of sets (or collections of sets) has to be dealt with carefully, however. To see that this is so, let us consider the following paradox, which is usually attributed to Bertrand Russell.

Consider the set of all sets. Some sets may contain themselves as elements while others may not. Most sets do not contain themselves. For example, the set of natural numbers does not contain itself; $\mathbb{N}$ is not an element of $\mathbb{N}$ since every element of $\mathbb{N}$ is a number and not a set of numbers. You may ask what a number is; Chapter 9 blurs the distinction between numbers and sets.

For the set of all sets, it must contain itself. There is no paradox here, yet.

Consider the set $\mathcal{R}$ of all sets that do not contain themselves as elements. We ask: is $\mathcal{R}$ contained in $\mathcal{R}$? If $\mathcal{R}$ is an element of $\mathcal{R}$, then, by definition of $\mathcal{R}$, $\mathcal{R}$ cannot contain $\mathcal{R}$. This is a contradiction. In case you think we were unwise to start with the assumption that $\mathcal{R}$ is an element of $\mathcal{R}$, consider the other case. If $\mathcal{R}$ is not an element of $\mathcal{R}$, then, again by the definition of $\mathcal{R}$, $\mathcal{R}$ is an element of $\mathcal{R}$. Again this is a contradiction.

Much effort was spent by mathematicians in trying to resolve this paradox. The usual remedy is to use axioms that do not allow the existence of the set of all sets. One axiom allows for the specification of subsets as we described in the introduction. For example, set difference is allowed by this axiom. Other axioms allow for unions, intersections, and

4.7. Russell's Paradox

power sets. That is, such a formal system of axioms allows us to do what we have done in this chapter.

If this all strikes you as too abstract, consider the following amusing version of Russell's paradox. If the barber in a village shaves all those, and only those, who do not shave themselves, then who shaves the barber?

4.8 Supplemental Exercises

Definition Review. Recall that *set*, *element*, and *is an element of* are undefined. There were thirteen italicized terms defined in this chapter. Define each of the following and give an example of each:

Definition 3. *A* is *empty*

Definition 4. *A* is a *subset* of *B*

Definition 11. *A* equals *B*

Definition 17. the *intersection* of *A* and *B*

Definition 21. *A* and *B* are *disjoint*

Definition 22. the *union* of *A* and *B*

Definition 29. the *intersection* of the sets in the collection of sets $\mathcal{A}$

Definition 30. $\mathcal{A}$ is a *pairwise disjoint* collection of sets

Definition 31. $\mathcal{A}$ is a *disjoint* collection of sets

Definition 35. the *union* of the sets in the collection of sets $\mathcal{A}$

Definition 39. the *difference* of *B* from *A*

Definition 44. the *complement* of *A*

Definition 45. the *power set* of *A*

Supplemental Exercise 1. List the elements of each of the following four sets.

$$A = \left\{ n \in \mathbb{N} \,\middle|\, \frac{12}{n} \in \mathbb{N} \right\}$$

$$B = \left\{ (x, y) \,\middle|\, x \in \mathbb{N}, y \in \mathbb{N}, x + y = 5 \right\}$$

$$C = \left\{ (x, y) \in \mathbb{R}^2 \,\middle|\, |x + y| \leq 0 \right\}$$

$$D = \left\{ x \in \mathbb{Z} \,\middle|\, \frac{x+2}{x-3} < 0 \right\}$$

Supplemental Exercise 2. After Definition 3, we remarked that $\{\{\}\}$ and $\{\emptyset\}$ are not empty. Verify this by finding an element of each. Are these sets equal? Explain!

Supplemental Exercise 3. Let *A* and *B* be sets. Can $A \cap B \supsetneq A$? Can $A \cap B \subsetneq B$? Explain.

Supplemental Exercise 4. Prove that $\{x^3 \mid x \in \mathbb{R}\} = \mathbb{R}$.

Supplemental Exercise 5. Determine which of the following sets are equal, and find those

4.8. Supplemental Exercises

that are are proper subsets of another.

$$A = \left\{x \in \mathbb{R} \mid \sqrt{x^2} = x\right\}$$

$$B = \left\{x \in \mathbb{R} \mid \frac{1+\sqrt{x}}{1-\sqrt{x}} \in \mathbb{R}\right\}$$

$$C = \left\{x \in \mathbb{R} \mid x \geq 0\right\}$$

$$D = \left\{x \in \mathbb{R} \mid \frac{x^3}{(x-1)^2} \in \mathbb{R}\right\}$$

$$E = \left\{x^2 \mid x \in \mathbb{R}\right\}$$

Supplemental Exercise 6. Prove or disprove that if $A \neq B$ and $B \neq C$, then $A \neq C$.

Supplemental Exercise 7. Prove that if $A \subset B$, $B \subset C$, $C \subset D$, and $D \subset A$, then $A = B = C = D$.

Supplemental Exercise 8. Let

$$A = \left\{x \in \mathbb{R} \mid x = \frac{a}{b} + \frac{b}{a}, \text{ for some } a, b \in \mathbb{R}^*\right\}$$

and

$$B = \left\{y \in \mathbb{R} \mid y = \sin t, \text{ for some } t \in \mathbb{R}\right\}.$$

Prove that $A \cap B = \emptyset$.

Supplemental Exercise 9. Find a condition so that $A \cap C = B \cap C$ implies $A = B$. Find an example where $A \cap C = B \cap C$ but $A \neq B$.

Supplemental Exercise 10. Let $A = \{1, 2, 3, 4, 5, 6\}$ and $B = \{2, 4, 6, 8\}$. Find all the subsets of $A \cap B$. Find all the subsets of $A \cup B$, containing exactly three elements.

Supplemental Exercise 11. Let $A = \{4, 5\}$. Suppose that $A \cup B = \{1, 2, 3, 4, 5, 8, 9, 10\}$ and $B \cup C = \{4, 5, 6, 7, 8, 9, 10\}$. Can we determine sets B and C such that this is true? Explain.

Supplemental Exercise 12. Let

$$A = \{n \in \mathbb{N} \mid n^2 - 8n + 7 > 0\},$$
$$B = \{x \in \mathbb{N} \mid x \text{ is a prime number}\},$$

and

$$C = \{n^2 \mid n \in \mathbb{N}\}.$$

Determine the sets $A \cap B$, $A \cap C$, $B \cup C$, $(A \cap B) \cup (A \cap C)$, and $(A \cap C) - (B \cup C)$.

Supplemental Exercise 13. Let A and B be sets. Prove that $A \cup B \subset B$ if and only if $A \subset B$. Conclude that $A \cup B = B$ if and only if $A \subset B$.

Supplemental Exercise 14. Let A and B be sets. Prove that $A \cap B = A \cup B$ if and only if $A = B$.

Supplemental Exercise 15. Prove the commutative property for unions. That is, for sets A and B show that $A \cup B = B \cup A$.

Supplemental Exercise 16. Prove the associative properties for intersections and unions. That is, for sets A, B, and C show that

$$(A \cap B) \cap C = A \cap (B \cap C) \quad \text{and} \quad (A \cup B) \cup C = A \cup (B \cup C).$$

Supplemental Exercise 17. Prove the distributive properties. That is, for sets A, B, and C show that

$$A \cap (B \cup C) = (A \cap B) \cup (A \cap C) \quad \text{and} \quad A \cup (B \cap C) = (A \cup B) \cap (A \cup C).$$

Supplemental Exercise 18. Let $\mathcal{A}$ be a collection of sets and let B be a set. Prove:

$$B \cap \left(\bigcup_{A \in \mathcal{A}} A \right) = \bigcup_{A \in \mathcal{A}} (B \cap A) \quad \text{and} \quad B \cup \left(\bigcap_{A \in \mathcal{A}} A \right) = \bigcap_{A \in \mathcal{A}} (B \cup A).$$

Supplemental Exercise 19. Find a condition so that $A \cup C = B \cup C$ implies $A = B$. Prove it. Find an example where $A \cup C = B \cup C$ but $A \neq B$.

Supplemental Exercise 20. Prove that $A \cup B = \emptyset$ if and only if $A = B = \emptyset$.

Supplemental Exercise 21. Prove that $A \cup (A \cap B) = A$.

Supplemental Exercise 22. Prove that if $C \subset A$, then $(A \cap B) \cup C = A \cap (B \cup C)$. Is the converse true?

Supplemental Exercise 23. Prove or disprove that if $C \subset B$, then

$$(A \cup B) \cap C = A \cup (B \cap C).$$

Supplemental Exercise 24. Let $A_1, \ldots, A_n$ be n sets. Prove that for $1 \leq i, j \leq n$,

$$A_i \cap A_j \supset A_1 \cap \cdots \cup A_n$$

and

$$A_i \cup A_j \subset A_1 \cup \cdots \cup A_n.$$

Supplemental Exercise 25. Let

$$\mathcal{A} = \{\{0, 1\}, \{1, 2\}, \{2, 3\}\},$$
$$\mathcal{B} = \{(0, 1), (1, 2), (2, 3)\},$$

and

$$\mathcal{C} = \{[0, 2], [1, 3], [2, 4]\}.$$

Which of $\mathcal{A}$, $\mathcal{B}$, and $\mathcal{C}$ are disjoint? Which are pairwise disjoint?

4.8. Supplemental Exercises

Supplemental Exercise 26. Let $\mathcal{C}$ be a collection of sets. Prove that if $\mathcal{C}$ contains exactly two sets, then $\mathcal{C}$ is disjoint if and only if $\mathcal{C}$ is pairwise disjoint.

Supplemental Exercise 27. Let $\mathcal{C}$ be a nonempty collection of sets. Prove that if $S \in \mathcal{C}$, then
$$\cap \mathcal{C} \subset S \subset \cup \mathcal{C}.$$

Supplemental Exercise 28. Let $\mathcal{C}$ be a collection of sets. Prove that $\cap \mathcal{C} = \cup \mathcal{C}$ if and only if $S_1 = S_2$ for all $S_1, S_2 \in \mathcal{C}$.

Supplemental Exercise 29. Give a counterexample to the converse of Proposition 34.

Supplemental Exercise 30. Prove that $A - B$ and $A \cap B$ are disjoint and
$$(A - B) \cup (A \cap B) = A$$
for sets A and B.

Supplemental Exercise 31. Let A, B, and C be sets. Prove that
$$A - (B \cap C) = (A - B) \cup (A - C) \text{ and } A - (B \cup C) = (A - B) \cap (A - C).$$
These properties are referred to as De Morgan's Laws.

Supplemental Exercise 32. Let A be a set and let $\mathcal{B}$ be a collection of sets. Prove that
$$A - \left(\bigcap_{B \in \mathcal{B}} B \right) = \bigcup_{B \in \mathcal{B}} (A - B) \quad \text{and} \quad A - \left(\bigcup_{B \in \mathcal{B}} B \right) = \bigcap_{B \in \mathcal{B}} (A - B).$$
These are also called De Morgan's Laws.

Supplemental Exercise 33. Suppose that we are given an increasing nested sequence of sets,
$$A_1 \subset A_2 \subset A_3 \subset \cdots.$$
Determine the intersection
$$\bigcap_{n=1}^{\infty} A_n.$$
Prove your answer. (Compare with Exercise 33.)

Supplemental Exercise 34. Suppose we are given the following decreasing nested sequence of sets,
$$B_1 \supset B_2 \supset B_3 \supset \cdots.$$
Determine the union
$$\bigcup_{n=1}^{\infty} B_n.$$
Prove your answer. (Compare with Exercise 36.)

Supplemental Exercise 35. Prove that $A - (A - B) = A \cap B$.

Supplemental Exercise 36. Prove that $A \cap (B - C) = (A \cap B) - (A \cap C)$.

Supplemental Exercise 37. Prove that $(A - B) \cup (B - A) = (A \cup B) - (A \cap B)$.

Supplemental Exercise 38. If $A - B = B - A$, what can you conclude about the two sets? Prove your answer. Answer the same question if $A - B \subset B - A$.

Supplemental Exercise 39. Prove that if $A \cup B = A \cap B$, then $A - B = \emptyset$.

Supplemental Exercise 40. Prove that $(A - B) \cap (C - A) = \emptyset$.

Supplemental Exercise 41. Prove that the following are equivalent conditions:
$A - B = B - A$,
$A = B$, and
$A - B = B - A = \emptyset$.

Supplemental Exercise 42. Prove that $(A')' = A$.

Supplemental Exercise 43. If $A = \{x \in \mathbb{R} \mid x^2 > x\}$, express A' as an interval.

Supplemental Exercise 44. Suppose that A and B are subsets of a universal set U. Prove that
$$A - B = A \cap B'.$$

Supplemental Exercise 45. Suppose that A and B are subsets of a universal set U. Prove that
$$(B')' = B \quad \text{and} \quad A' - B' = B - A.$$

Supplemental Exercise 46. Suppose that A and B are subsets of a universal set U. Prove that
$$(A - B) \cup (B - A) = (A \cap B)' - (A \cup B)'.$$

The set $(A - B) \cup (B - A)$ is sometimes called the *symmetric difference of A and B*.

Supplemental Exercise 47. Prove that $B \subset (A \cap B')'$.

Supplemental Exercise 48. Let A and B be sets such that $\mathcal{P}(A) = \mathcal{P}(B)$. Prove that $A = B$.

Supplemental Exercise 49. Show that $(A \cap B \cap C)' = A' \cup B' \cup C'$. Is it true that $(A \cup B \cup C)' = A' \cap B' \cap C'$?

Supplemental Exercise 50. Prove that if $(A \cup B)' = A' \cup B'$, then $A = B$. Does $(A \cap B)' = A' \cap B'$ imply that $A = B$?

Supplemental Exercise 51. Prove that $(A' \cup B')' \cup (A' \cup B)' = A$.

Supplemental Exercise 52. Can there exist a set whose power set has $4^k + 1$ elements, where $k \in \mathbb{N}$?

Supplemental Exercise 53. Prove that $(A - B) \cap (B - A) = \emptyset$ for all sets A and B.

Supplemental Exercise 54. Let A and B be sets. Prove or disprove by a counterexample:
$$\mathcal{P}(A) \cap \mathcal{P}(B) \subset \mathcal{P}(A \cap B)$$

4.8. Supplemental Exercises

and

$$\mathcal{P}(A) \cap \mathcal{P}(B) \supset \mathcal{P}(A \cap B).$$

Supplemental Exercise 55. Let A and B be sets. Prove or disprove by a counterexample:

$$\mathcal{P}(A) \cup \mathcal{P}(B) \subset \mathcal{P}(A \cup B)$$

and

$$\mathcal{P}(A) \cup \mathcal{P}(B) \supset \mathcal{P}(A \cup B).$$

Supplemental Exercise 56. Let A and B be sets. Prove or disprove by a counterexample:

$$\mathcal{P}(A) - \mathcal{P}(B) \subset \mathcal{P}(A - B)$$

and

$$\mathcal{P}(A) - \mathcal{P}(B) \supset \mathcal{P}(A - B).$$

Supplemental Exercise 57. Prove that $\mathcal{P}(A - B) \subset (\mathcal{P}(A) - \mathcal{P}(B)) \cup \{\emptyset\}$.

Supplemental Exercise 58. Project. We have taken a relatively naive approach to set theory. It is possible to be more axiomatic. Find a text on axiomatic set theory and outline the development of the topics of this chapter axiomatically, giving undefined terms, axioms, definitions, etc. Pay particular attention to the axioms that allow new sets to be created from known sets, such as by unions and intersections. Is a universal set needed in your axiomatic set theory? How is Russell's Paradox avoided in your axiomatic set theory?

5

Functions

Philosophy is written in this grand book—I mean the universe—which stands continually open to our gaze, but it cannot be understood unless one first learns to comprehend the language and interpret the characters in which it is written. It is written in the language of mathematics, and its characters are triangles, circles, and other geometrical figures, without which it is humanly impossible to understand a single word of it; without these, one is wandering about in a dark labyrinth.

— Galileo Galilei [4]

5.1 Functions as Rules

Galileo's words reflect a qualitative differentiation with respect to how scientific endeavor was thought to be properly conducted before the seventeenth century. With few exceptions, of which the most notable was the work of Archimedes, before the advent of modern science, scientific inquiry was under the influence of Greek philosophy that focused on discovering the causes and reasons that were behind the occurrence of natural phenomena. The decisive break with this tradition was exemplified in the effort to shift the emphasis towards giving a quantitative description of phenomena. The following historical example, related to the phenomenon of a free falling object, illustrates this method that gave rise to the notion of *function*, which is perhaps the single most important concept in all of mathematics.

Suppose a ball is dropped from a height. What questions can we ask? A Greek philosopher would be concerned with the reason why the object falls. Aristotle would say that an object tends to occupy its natural position, which is being at rest. Galileo singled out two variable quantities he deemed pertinent for the study of this phenomenon and sought to discover a relationship between them. Galileo's question was "what is the distance of the ball from its starting position as time goes by?" The table below gives us an approximation of the values obtained in this experiment, where time is measured in seconds and distance is measured in meters.

Can you guess what number corresponds to 6 if the pattern continues? Can you see the pattern? We can discern a pattern from the second column. But what if $t = \sqrt{2}$? In general, how are s and t related? Galileo was able to see that $s = 5t^2$, which is a good experimental observation. You may recall the equation $s = \frac{1}{2}gt^2$ from physics where g denotes acceleration due to gravity. The value of g averages around 9.8 m/sec^2. Without

Time (t)	Distance (s)
0	0
1	5
2	20
3	45
4	80
5	125
6	?

mechanical timing, Galileo's estimate of 10 m/sec^2, yielding the formula $s = \frac{1}{2}gt^2 = 5t^2$, is quite good.

The formula $s = 5t^2$ is an example of a quantitative description of a natural phenomenon. It represents a functional relationship between two variable quantities and is the precursor of the formal mathematical definition of a function.

The table might remind you of an informal definition of "function." Probably you have seen (but perhaps forgotten) a definition of "function" that goes like this: a function is a pair of sets and a rule that assigns a unique element of the second set to each element of the first set. You may also recall that the first set is called the domain of the function and you may recall (somewhat incorrectly!) that the second set is called the range of the function. Actually, the second set should be called the codomain of the function. Probably your facility with functions is better than your recollection of the definition of the term.

In the next section, we will define functions by sets. From a rigorous (logical, axiomatic) point of view, this is preferable. However, you may (or should) continue to think of a function as a rule that assigns to each value of its domain a unique value in its range.

Before you continue reading, think for a moment about a function and its graph. If, for a graph, you are thinking of a pictorial representation, then what does that picture represent? Is there a difference between that object and the function itself?

5.2 Cartesian Products, Relations, and Functions

Before giving our definition of the word function, we will define the cartesian product of two sets, often simply called a product.

Definition 1. Let A and B be sets. The *cartesian product* of A and B, denoted by $A \times B$, is the set of all ordered pairs (a, b) such that $a \in A$ and $b \in B$.

While it is possible to define the term *ordered pair* by something like

$$(a, b) = \{a, \{a, b\}\},$$

you may accept this as undefined. Since the notation (a, b) also represents an open interval, some authors use the notation $a \times b$ for an ordered pair.

You should have in mind the usual example and its model. $\mathbb{R} \times \mathbb{R}$, which is usually denoted as $\mathbb{R}^2$, is the setting for most functions you have studied; the model of $\mathbb{R}^2$ is the plane that you usually draw by indicating the two coordinate axes. A point P in it corresponds to an ordered pair (x, y); the vertical line through P intersects the horizontal axis at x and the horizontal line through P intersects the vertical axis at y. See Figure 5.1.

5.2. Cartesian Products, Relations, and Functions

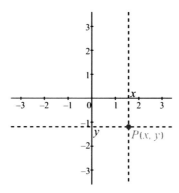

Figure 5.1. A point of the plane models an ordered pair (x, y) in $\mathbb{R}^2$

It is possible to construct the products of other sets. Using the geometric intuition that we used for $\mathbb{R}^2$, we may model products. For example, $\mathbb{R}^3 = \mathbb{R} \times \mathbb{R} \times \mathbb{R}$ is modeled by (three-dimensional) space.

You may wonder if $\mathbb{R} \times \mathbb{R} \times \mathbb{R}$ means $(\mathbb{R} \times \mathbb{R}) \times \mathbb{R}$ or $\mathbb{R} \times (\mathbb{R} \times \mathbb{R})$. It does not matter. The reason for this has to do with the concept of *isomorphism*, which we will introduce in Chapter 8. The idea is that every element $((x, y), z) \in (\mathbb{R} \times \mathbb{R}) \times \mathbb{R}$ corresponds to exactly one element $(x, (y, z)) \in \mathbb{R} \times (\mathbb{R} \times \mathbb{R})$ and we can think of both as $(x, y, z) \in \mathbb{R}^3$. We will give a different definition of products of collections of sets in Definition 33 of Chapter 7.

Similarly, $\mathbb{N}^2 = \mathbb{N} \times \mathbb{N}$ is modeled by a lattice of isolated points and $\mathbb{R} \times \{1, 2\}$ and $\{1, 2\} \times \mathbb{R}$ are modeled by pairs of lines. We draw pictures of these models as if to represent the sets themselves.

Exercise 2. Draw pictures of (the models of) $\mathbb{R}^3$, $\mathbb{N}^2$, $\mathbb{R} \times \{1, 2\}$, and $\{1, 2\} \times \mathbb{R}$.

We did not require non-empty sets in Definition 1. The product with the empty set is empty.

Exercise 3. Let A and B be sets. Prove that $A \times B = \emptyset$ if and only if $A = \emptyset$ or $B = \emptyset$.

The operation of taking products is not commutative. For example, if $A = \{1\}$ and $B = \{2\}$, then $A \times B = \{(1, 2)\}$ while $B \times A = \{(2, 1)\}$. The following exercise relates to this.

Exercise 4. Let A and B be sets. Prove that $A \times B$ and $B \times A$ are disjoint if and only if A and B are disjoint.

Before giving a definition of the term function, we define a simpler term: relation.

Definition 5. Let A and B be sets. A *relation* from A to B is a subset

$$R \subset A \times B.$$

This relation is denoted as $R(A, B)$ or, when A and B are understood, simply as R. If $A = B$, then $R \subset A \times A = A^2$ and we call R a *relation* on A.

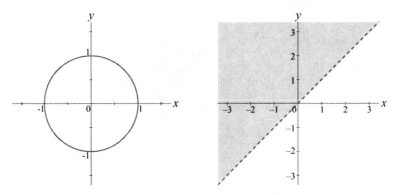

Figure 5.2. The circle $x^2 + y^2 = 1$ in $\mathbb{R}^2$ and $<$ on $\mathbb{R}$

An ordered pair (a, b) in a relation R is also denoted by $a \, R \, b$, which is read "a is R-related to b" or as "a is related to b." If we write $a \, R \, b$, we mean $(a, b) \in R$ and $a \, \not{R} \, b$ will mean that $(a, b) \notin R$.

Let us look at two examples of relations that you have seen before. They are represented by Figure 5.2.

Example 6. The equation $x^2 + y^2 = 1$ determines a subset of $\mathbb{R}^2$, that is, a relation on $\mathbb{R}$.

Similarly, any equation in two variables determines a relation.

Example 7. We define a relation $< (\mathbb{R}, \mathbb{R})$, in the notation of the definition. This is the usual relation "less than" on the set of real numbers. We use the notation $x < y$ to mean that (x, y) is an element of this relation. We will refer to this as the relation $<$ on $\mathbb{R}$ and never use the hideous notation $< (\mathbb{R}, \mathbb{R})$ again.

In Chapter 6, we will study relations on a set more closely.

For the remainder of this chapter, we will mostly work with functions. Every function is a relation. However, the converse is false: the circle of Example 6 is not the equation of a function. No doubt you would suspect that the relation of Example 7, as depicted in Figure 5.2, is not a function either.

We are now ready to give a formal definition of function. It will be different from the informal definition in Section 5.1. In fact, the object we define is often called the *graph* of a function rather than a function.

Definition 8. Let X and Y be sets. A *function* from X to Y is a relation f from X to Y with the property that for every $x \in X$ there exists a unique $y \in Y$ such that the ordered pair (x, y) is in f. X is called the *domain* of the function and Y is called the *codomain* of the function. We call the set f the *graph* of f.

Remark 9. Definition 8 did not define "function," but defined "function from a set X to a set Y." Therefore, we could define a function as an ordered triple,

$$(X, Y, f).$$

It consists of three sets: X is the domain of the function, Y is the codomain of the function, and f is the graph of the function, with $f \subset X \times Y$ such that for every $x \in X$ there exists

5.2. Cartesian Products, Relations, and Functions

a unique $y \in Y$ such that (x, y) is in f. In fact, the domain can be implied by the set f of ordered pairs. See Supplemental Exercise 18.

That is, a function f from X to Y is a subset of $X \times Y$ such that for every $x \in X$ there exists a unique $y \in Y$ with $(x, y) \in f$. When we sketch the graph of a function, the domain is usually on the horizontal axis while the codomain is on the vertical axis.

The circle of Example 6 is a relation on $\mathbb{R}$. It fails to define a function from $\mathbb{R}$ to $\mathbb{R}$ in two ways. First, the domain is not $\mathbb{R}$ since x must be between -1 and 1, inclusive. Second, domain values do not, in general, correspond to unique codomain values. For example, $(3/5, 4/5)$ and $(3/5, -4/5)$ are both on the circle. So the circle is not a function with domain $[-1, 1]$ either.

Remark 10. Here are some important words about notation. A function, as in Definition 8, is denoted by $f : X \to Y$. When X and Y are understood, we may simply denote the function by f. If
$$(x, y) \in f,$$
then we denote the element y by $f(x)$; that is, $f(x) = y$ and $(x, y) = (x, f(x))$. While most mathematicians are happy to call a function f, many frown at calling a function $f(x)$ since $f(x)$ is an element of the codomain. The notation $y = f(x)$ is better and is commonly used, as in the function $y = x^2$ or $f(x) = x^2$.

We used a lower case f to represent a set but there is no rule that requires the name of a set to be an upper case Latin letter. Some authors use the words *source* and *target* for domain and codomain, respectively. When specifying a function f without explicitly giving the domain and codomain, we will use the notations $\mathcal{D}_f$ and $\mathcal{C}_f$ for the domain and codomain, respectively.

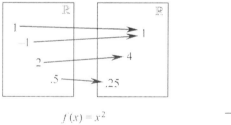
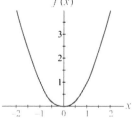

Figure 5.3. A function thought of as a rule and as a subset of $\mathbb{R}^2$

The formal definition looks different from the informal definition, but it is not so different at all. The domain and codomain are the same in both definitions. The difference is in the interpretation of f, which is either a rule or a set. The primary difference may be in the way one thinks of a function. However, we ignore this ambiguity and think of a function in whatever way is convenient at the time; see Figure 5.3.

So what happened to the range?

Definition 11. Let $f : X \to Y$ be a function and $S \subset X$. The *image* of S under f is the set
$$f(S) = \{ y \in Y \mid (x, y) \in f \text{ for some } x \in S \}.$$

The following gives us another way to think of the image of a set.

Exercise 12. Let $f : X \to Y$ be a function and $S \subset X$. Prove that
$$f(S) = \{f(x) \mid x \in S\}.$$

Finally, the range can be described as the image of the domain.

Definition 13. The *range* of a function $f : X \to Y$ is the image of the domain $f(X)$.

The following examples may add clarity.

Example 14. Figure 5.3 shows a picture of the function $f(x) = x^2$. What are the domain and codomain for this function? The domain is $\mathbb{R}$. By convention, when the domain is not given explicitly, then it is the largest subset of real numbers (or whatever set fits the situation) for which the rule defining the function makes sense. The range of the function is $[0, \infty)$. Since the codomain must contain the range, any set containing the set $[0, \infty)$ would suffice. For real-valued functions, that is, functions whose range is a subset of $\mathbb{R}$, we use the convention that the codomain is $\mathbb{R}$. That is, we can write $f : \mathbb{R} \to \mathbb{R}$. What is f? Since the domain of f is $\mathbb{R}$ and the codomain is $\mathbb{R}$, the graph is a subset of $\mathbb{R}^2$:
$$f = \{(x, x^2) \mid x \in \mathbb{R}\}.$$

Example 15. The domain and range of the function $f(x) = \sqrt{x}$ is $[0, \infty)$ and its codomain is $\mathbb{R}$. We can write $f : [0, \infty) \to \mathbb{R}$. Its graph is
$$f = \{(x, \sqrt{x}) \mid x \in [0, \infty)\}.$$

Exercise 16. Draw a picture of f from Example 15.

In essence, if we know the rule defining a function f and its domain X, then we know that
$$f = \{(x, f(x)) \mid x \in X\}.$$
We did not specify the codomain of the function f. Hence, we do not know the product that appears in the definition of function that is supposed to contain the set f. However, this is generally good enough. In practice, we may be given a formula defining a function where the domain and range are implied and the codomain may be inferred.

Exercise 17. Determine the domain, range, and function defined by the rule $y = \frac{x}{x^2-1}$. Draw a picture of f.

When are two functions equal? Functions depend on three sets: the domain X, the codomain Y, and the set of ordered pairs in $X \times Y$. It makes sense to say two functions are equal if their domains, codomains, and sets of ordered pairs are equal.

Definition 18. The functions $f : X \to Y$ and $g : U \to V$ are *equal* if and only if $X = U$, $Y = V$, and, as sets of ordered pairs, $f = g$.

5.2. Cartesian Products, Relations, and Functions

In practice, to check that two functions f and g are equal, we check that $f(x) = g(x)$ for all x. You should ask yourself "For all x in what set?" In fact, one must first check that the domains of f and g are equal and then that $f(x) = g(x)$ for all $x \in \mathcal{D}_f = \mathcal{D}_g$; that is, you must check that $X = U$ in the notation of Definition 18.

Example 19. Let
$$f : \mathbb{R} \to \mathbb{R} \quad \text{and} \quad g : \mathbb{R}_+ \to \mathbb{R}$$
be defined by
$$f(x) = x \quad \text{and} \quad g(x) = x$$
for all x in the respective domains. These are different functions: they are not equal since their domains are different. However, $f(x) = g(x)$ for all $x \in \mathbb{R}_+ = \mathcal{D}_g$. On the other hand, $f(-1) = -1$ while $g(-1)$ is not defined since $-1 \notin \mathbb{R}_+ = \mathcal{D}_g$. Another way of saying this is that $(-1,-1) \in f$ but $(-1,-1) \notin g$.

In Example 19, the functions f and g are similar. You can think of g as being f with a restricted domain. This idea is important enough to merit a name and some notation.

Definition 20. Let $f : X \to Y$ be a function and let $S \subset X$. The function $g : S \to Y$ defined by $g(x) = f(x)$ for all $x \in S$ is called the *restriction* of f to S and is denoted $f|_S$.

We continue with exercises and an example about the image of sets under a function.

Exercise 21. Let f be a function and let $A \subset B \subset \mathcal{D}_f$. Prove that
$$f(A) \subset f(B).$$

Exercise 22. Let f be a function and let A and B be subsets of $\mathcal{D}_f$. Prove that
$$f(A \cup B) = f(A) \cup f(B).$$

Intersections are more complicated.

Example 23. Let $f : \mathbb{R} \to \mathbb{R}$ be defined by $f(x) = |x|$ and let $A = (-2,-1)$ and $B = (1,2)$ be intervals. Since $A \cap B = \emptyset$, $f(A \cap B) = \emptyset$. Also, $f(A) \cap f(B) = (1,2)$. Therefore,
$$f(A \cap B) \neq f(A) \cap f(B).$$

Another example where $f(A \cap B) \neq f(A) \cap f(B)$ is $f : \mathbb{R} \to \mathbb{R}$ defined by $f(x) = x^2$ with $A = \{-1\}$ and $B = \{1\}$. Then $A \cap B = \emptyset$, but $f(A) = f(B) = \{1\}$.
In both cases, $f(A \cap B) \subsetneq f(A) \cap f(B)$.

Containment in one direction always holds for intersections.

Exercise 24. Let f be a function and let A and B be subsets of $\mathcal{D}_f$. Prove that
$$f(A \cap B) \subset f(A) \cap f(B).$$

5.3 Injective, Surjective, and Bijective Functions

In this section we will look at some properties of functions that may be familiar.

Definition 25. A function $f : X \to Y$ is *one-to-one* if and only if, for all $x_1, x_2 \in X$, $x_1 \neq x_2$ implies $f(x_1) \neq f(x_2)$.

For a one-to-one function we often write that it as "1-1." A one-to-one function is also called an *injective* function or an *injection*.

The following exercise gives us an equivalent definition of one-to-one.

> **Exercise 26.** Prove that a function $f : X \to Y$ is one-to-one if and only if, for all x_1 and x_2 in X, $f(x_1) = f(x_2)$ implies $x_1 = x_2$.

One-to-one functions are better behaved than generic functions. Compare Exercises 22, 24, and Supplemental Exercise 20.

Definition 27. A function $f : X \to Y$ is *onto* if and only if $f(X) = Y$, that is, the range of f is the codomain of f.

An onto function is also called a *surjective* function or a *surjection*. The following exercise gives us an equivalent definition of onto.

> **Exercise 28.** Prove that a function $f : X \to Y$ is onto if and only if, for each $y \in Y$, there exists $x \in X$ such that $f(x) = y$.

Definition 29. A function $f : X \to Y$ is *bijective* if and only if it is one-to-one and onto.

Bijective functions are also called *one-to-one correspondences*. We generally prefer "bijective" function or "bijection" to "one-to-one correspondence;" however, we prefer the latter for cardinality, the topic of Chapter 7.

The following will remind you of the tenuous role of the codomain as compared with the range.

Example 30. Let us show that for every function $f : X \to Y$, there exists a surjective function from X onto $f(X)$. Define $g : X \to f(X)$ by $g(x) = f(x)$ for all $x \in X$. g is onto by design:

$$g(X) = \{g(x) \mid x \in X\} = \{f(x) \mid x \in X\} = f(X).$$

> **Exercise 31.** Prove that if $f : X \to Y$ is an one-to-one function, then there exists a bijective function from X onto $f(X)$.

The following exercise gives us an equivalent definition of bijective.

> **Exercise 32.** Prove that a function $f : X \to Y$ is bijective if and only if, for each $y \in Y$, there exists a unique $x \in X$ such that $f(x) = y$.

It is possible for a function to be bijective, to have exactly one of the properties of one-to-one and onto, or to have neither. For example, $f(x) = x(x-1)(x-2)$, with the conventional codomain of $\mathbb{R}$ is onto, but not one-to-one since $f(0) = f(1) = f(2) = 0$.

5.4 Compositions of Functions

An important operation on functions is that of composition. The most common defi[n] requires that the codomain of the first function equals the domain of the second functi[on] in

$$f : X \to Y \quad \text{and} \quad g : Y \to Z.$$

Since the range of a function is more essential than its codomain, we allow the range of the first function to be a subset of the domain of the second function. Perhaps you recall defining the composite function $g \circ f : X \to Z$ by

$$(g \circ f)(x) = g(f(x)).$$

Below, we define the composition of two functions as a subset of a cartesian product. A more general definition, which does not require the range of the first function to be contained in the domain of the second function, is given in Section 5.6; it evolves from the definition of composition of relations.

Definition 33. Let $f : X \to Y$ and $g : W \to Z$ be functions with $f(X) \subset W$. The *composition* of f followed by g is the function, $g \circ f$, defined by

the domain of $g \circ f$ is X, the domain of f,
the codomain of $g \circ f$ is Z, the codomain of g, and
the graph of $g \circ f$ is

$$\{(x, z) \in X \times Z \mid (x, y) \in f \text{ and } (y, z) \in g \text{ for some } y \in f(X)\}.$$

Definition 33 is reasonable. The idea is that f sends an $x \in X$ to some $y \in Y$. To be able to apply g, we must have $y \in W$ also; since $f(X) \subset W$ this is automatically satisfied. Finally, g sends y to some $z \in Z$. If $Y \subset W$ (for example, when $Y = W$), then $f(X) \subset W$.

Exercise 34. For $f : X \to Y$ and $g : W \to Z$ with $f(X) \subset W$, verify that

$$g \circ f = \{(x, g(f(x))) \mid x \in X\}.$$

Given two functions f and g, there are often two composite functions: $f \circ g$ and $g \circ f$. However, in general,

$$f \circ g \neq g \circ f,$$

as the following examples demonstrate. This is true even when both composite functions exist.

Example 35. Define $f : \mathbb{R} \to \mathbb{R}$ and $g : \mathbb{R} \to \mathbb{R}$ by $f(x) = 2x$ and $g(x) = 2x + 1$ for all $x \in \mathbb{R}$. Then $f \circ g$ and $g \circ f$ are both functions from $\mathbb{R}$ to $\mathbb{R}$. However,

$$(f \circ g)(x) = f(g(x)) = f(2x + 1) = 2(2x + 1) = 4x + 2$$

while

$$(g \circ f)(x) = g(f(x)) = g(2x) = 2(2x) + 1 = 4x + 1.$$

So $f \circ g \neq g \circ f$. Therefore, composition is not commutative.

The following example is more peculiar, yet more typical.

Example 36. Let $f(x) = 2 + \sqrt{x}$ and $g(x) = \sqrt{1-x}$. Then,

$$f(g(x)) = 2 + \sqrt{\sqrt{1-x}} = 2 + \sqrt[4]{1-x}$$

and hence

$$f \circ g = \left\{ (x, 2 + \sqrt[4]{1-x}) \mid x \in (-\infty, 1] \right\}.$$

The range of g equals the domain of f, the interval $[0, \infty)$. This is consistent with Definition 33.

On the other hand,

$$g(f(x)) = \sqrt{1 - (2 + \sqrt{x})} = \sqrt{-1 - \sqrt{x}}$$

is not defined for any real number x. Hence, $g \circ f$ does not exist. Another way of saying this is that $g \circ f$ does not exist since the domain of g, which is $(-\infty, 1]$, does not contain the range of f, which is $[2, +\infty)$. The domain of g and the range of f are disjoint sets. So, $f \circ g \neq g \circ f$.

Examples 35 and 36 show that composition depends on the order of the functions, so it is not commutative. The second example also shows that one of the compositions may be more interesting than the other. Usually we will have the following situation: given $f : X \to Y$ and $g : W \to Z$, if $Y \subset W$ (or $Y = W$), then we have the composite $g \circ f : X \to Z$.

Though composition is not commutative, it is associative.

Exercise 37. Show that composition of functions is associative.

Compositions of one-to-one and onto functions are well-behaved.

Exercise 38. Prove that the composition of one-to-one functions is one-to-one.

Exercise 39. Prove that if $f : X \to Y$ and $g : Y \to Z$ are onto functions, then $g \circ f$ is onto.

In Definition 33, $g \circ f$ was defined whenever the range of f was a subset of the domain of g. In Exercise 39, we strengthened this requirement to equality: $Y = \mathcal{R}_f = \mathcal{D}_g$. The following example shows that this stronger requirement is necessary.

Example 40. Let $f : \mathbb{R}_+ \to \mathbb{R}_+$ and $g : \mathbb{R} \to \mathbb{R}$ be defined by $f(x) = x$ and $g(x) = x$ for all x in their respective domains. Both functions are onto. By Definition 33, $g \circ f$ has domain $\mathbb{R}_+$ and codomain $\mathbb{R}$; $g \circ f$ is not onto.

From Exercises 38 and 39 we see that the composite $g \circ f$ of two bijections $f : X \to Y$ and $g : Y \to Z$ is a bijection.

The converses of Exercises 38 and 39 are interesting and are considered in Supplemental Exercises 22 and 23.

5.5 Inverse Functions and Inverse Images of Functions

The study of inverses is common in mathematics. They require identities and composites.

Definition 41. A function f is the *identity* function on X if and only if $\mathcal{D}_f = \mathcal{C}_f = X$ and
$$f = \{(x, x) \mid x \in X\} \subset X \times X.$$

That is, $f : X \to X$, defined by $f(x) = x$ for all $x \in X$, is the identity function on X. We denote it by I_X.

For functions $f : X \to Y$ and $g : W \to X$, $f \circ I_X = f$ and $I_X \circ g = g$. See Supplemental Exercise 24

Theorem 42. *I_X is the unique function from X to X that satisfies $f \circ I_X = f$ and $I_X \circ g = g$ for all functions $f : X \to Y$ and $g : W \to X$.*

Proof. Let $i : X \to X$ be a function that satisfies $f \circ i = f$ and $i \circ g = g$ for all functions $f : X \to Y$ and $g : W \to X$. We see that

$$i = i \circ I_X \tag{1}$$
$$= I_X, \tag{2}$$

where (1) follows from the definition of I_X and (2) follows from the definition of i. □

Definition 43. Let f and g be functions. f and g are *inverse* functions if and only if
$$f \circ g = I_{\mathcal{D}_g} \quad \text{and} \quad g \circ f = I_{\mathcal{D}_f}.$$

We say that g is an *inverse* of f. A function f is *invertible* if and only if an inverse function g exists.

This definition implies a relationship between the domains and codomains of the inverse functions, as given in the following exercise.

Exercise 44. Prove that if f is an invertible function and g is an inverse of f, then
$$\mathcal{C}_g = \mathcal{D}_f \quad \text{and} \quad \mathcal{C}_f = \mathcal{D}_g.$$

It is important to include both $f \circ g = I_{\mathcal{D}_g}$ and $g \circ f = I_{\mathcal{D}_f}$ in the definition of inverse functions, as Example 45 will show.

Example 45. Let $f : \mathbb{R}_+ \to \mathbb{R}$ be defined by $f(x) = x$ and let $g : \mathbb{R} \to \mathbb{R}_+$ be defined by $g(x) = |x|$. Then $(g \circ f)(x) = g(x) = x$ for all $x \in \mathbb{R}_+$; that is, $g \circ f = I_{\mathbb{R}_+}$. However, $(f \circ g)(x) = |x|$ for all $x \in \mathbb{R}$, which is not $I_{\mathbb{R}}$. So, f and g are not inverse functions. We have f is one-to-one but not onto, and g is onto but not one-to-one.

Exercise 46. Prove that a function is invertible if and only if it is bijective.

If it sounds strange to read "an inverse" rather than "the inverse" in the definition of inverse functions, Theorem 47 will help. We will always write "the inverse" after the following exercise.

Theorem 47. *Every invertible function has a unique inverse function.*

Proof. Let $f : X \to Y$ be an invertible function. Suppose that g and h are inverses of f. By Exercise 44, g and h are functions from Y to X. We use the associative property to see that

$$h = I_X \circ h \qquad (3)$$
$$= (g \circ f) \circ h \qquad (4)$$
$$= g \circ (f \circ h)$$
$$= g \circ I_Y \qquad (5)$$
$$= g, \qquad (6)$$

where (3) and (6) follow from the definition of identity function, and (4) and (5) follow from the definition of inverse function. □

Since inverse functions are unique, when they exist, we have a notation for them: the inverse of f is denoted by f^{-1}.

When f is invertible, we have an inverse function f^{-1} and it makes sense to take images of the inverse function such as $f^{-1}(S)$. We call this the inverse image of S, which can be generalized to noninvertible functions.

Definition 48. Let $f : X \to Y$ be a function and $T \subset Y$. The *inverse image* of T under f is

$$f^{-1}(T) = \{x \in X \mid (x, y) \in f \text{ for some } y \in T\}.$$

Remark 49. Be careful! The inverse image of a set is a set, not an element. This is true even when the set is a singleton. When T is a singleton set $\{y\}$, we write $f^{-1}(y)$ instead of $f^{-1}(\{y\})$. Technically this is wrong, but it is an accepted case of "abuse of notation." Similarly, when $f^{-1}(T)$ is a singleton set $\{x\}$, we write $f^{-1}(T) = x$ instead of $f^{-1}(T) = \{x\}$.

Example 50. The use of the superscript in $f^{-1}(T)$ should not make you think the function f is invertible. For example, let $f : \mathbb{R} \to \mathbb{R}$ be defined by $f(x) = |x|$. Then

$$f^{-1}(0) = 0, \quad f^{-1}(1) = \{-1, 1\}, \quad \text{and} \quad f^{-1}(-1) = \emptyset.$$

We abused notation again:

$$f^{-1}(\{0\}) = \{0\}, \quad f^{-1}(\{1\}) = \{-1, 1\}, \quad \text{and} \quad f^{-1}(\{-1\}) = \emptyset.$$

The use of inverse images gives us a way of describing one-to-one and onto functions.

Exercise 51. Let $f : X \to Y$ be a function. Prove the following:
(a) f is one-to-one if and only if $f^{-1}(y)$ contains at most one element for every $y \in Y$.
(b) f is onto if and only if $f^{-1}(y)$ contains at least one element for every $y \in Y$.
(c) f is bijective if and only if $f^{-1}(y)$ contains exactly one element for every $y \in Y$.

5.6. Another Approach to Compositions

Supplemental Exercise 32 shows that inverse images behave nicely with respect to set operations.

Forward and inverse images can be combined. (See Supplemental Exercises 33 and 34.) The following exercises show that, in general, the new set is different from the original.

Exercise 52. Find a function $f : X \to Y$ and $S \subset X$ so that $f^{-1}(f(S)) \neq S$.

Exercise 53. Find a function $f : X \to Y$ and $T \subset Y$ so that $f(f^{-1}(T)) \neq T$.

5.6 Another Approach to Compositions

In Section 5.4 of Chapter 5, we looked at compositions of functions where the range of the first function was contained in the domain of the second function. In this section, we will remove this requirement. The moral of this section is that the easier (i.e., more general) the definitions are, the harder (i.e., less general) the theorems are. Relations can be composed.

Definition 54. Let f be a relation from X to Y and let g be a relation from W to Z. The *composition* of f followed by g is the relation from X to Z defined by

$$g \circ f = \{(x, z) \in X \times Z \mid \text{there exists } y \in Y \cap W \text{ such that } (x, y) \in f \text{ and } (y, z) \in g\}.$$

It is more common to compose functions than relations. However, we will use this idea to generalize composition of functions.

Definition 55. Let $f : X \to Y$ and $g : W \to Z$ be functions. The *composition* of f followed by g is the function defined by

$$g \circ f = \{(x, z) \in X \times Z \mid \text{there exists } y \in Y \cap W \text{ such that } f(x) = y \text{ and } g(y) = z\}.$$

The codomain of $g \circ f$ is Z, the codomain of g. The domain of $g \circ f$ is the largest subset of X for which it makes sense.

Definition 55 differs from Definition 33 in two ways: first, $f(X) \subset W$ is not required and, second, the domain is a subset of X, but not necessarily all of X. The domain may turn out to be the empty subset of X. The phrase "the largest subset of X for which it makes sense" may not seem very precise. An example should help.

Example 56. Let us reconsider Example 36. Let $f(x) = 2 + \sqrt{x}$ and $g(x) = \sqrt{1-x}$. Since $(g \circ f)(x) = \sqrt{1 - (2 + \sqrt{x})} = \sqrt{-1 - \sqrt{x}}$ is not defined for any real number x, the domain of $g \circ f$ is empty. Hence, $g \circ f = \emptyset$. This is possible since the definition of a function does not require a nonempty domain and $\emptyset \times Z = \emptyset$ for every set Z.

The following is similar.

Exercise 57. Suppose that $w, x, y,$ and z are distinct. Let $X = \{x\}$, $Y = \{y\}$, $W = \{w\}$, and $Z = \{z\}$. Define $f : X \to Y$ and $g : W \to Z$ by $f(x) = y$ and $g(w) = z$. Describe $g \circ f$ and $f \circ g$. Do f and g commute under composition? (That is, does $g \circ f$ equal $f \circ g$?)

In general, what is the domain of $g \circ f$? The composition of f followed by g may be rewritten as
$$g \circ f = \{(x, g(f(x))) \mid x \in \mathcal{D}_f \text{ and } f(x) \in \mathcal{D}_g\}.$$
This tells us that the domain is
$$\mathcal{D}_{g \circ f} = \{x \in \mathcal{D}_f \mid f(x) \in \mathcal{D}_g\}.$$

We have already seen (Examples 35, 36, and 56, along with Exercise 57) that composition is not commutative. Composition is still associative, but the proof requires careful consideration of the domains. This is an example of where our more general definition makes a theorem more difficult to prove.

Exercise 58. Show that composition of functions is associative.

Let us revisit the relationship of compositions with the properties of being one-to-one and onto. The following exercise gives us examples of how generalizing a definition can turn a true theorem into a false statement.

Exercise 59. For each of (a)–(d), give f and g satisfying the conditions:
(a) f and g are onto, but $g \circ f$ is not onto.
(b) neither f nor g is one-to-one, but $g \circ f$ is one-to-one.
(c) f is a bijection, g is not one-to-one, but $g \circ f$ is one-to-one.
(d) g is a bijection, f is not onto, but $g \circ f$ is onto.

Recall that the definition of inverses is dependent on composition of functions. You may wish to redo the exercises in Sections 5.4 and 5.5, this time using Definition 55 for the definition of composition of functions.

5.7 Supplemental Exercises

Definition Review. Twenty-eight italicized terms were defined in this chapter. Let A, B, X, Y, S, and T be sets with $S \subset X$ and $T \subset Y$. Let $f : X \to Y$ and $g : W \to Z$ be functions (or relations). Assume the sets are subsets of a universal set U. Define each of the following and give an example:

Definition 1. the *cartesian product* of A and B

Definition 5. a *relation* from A to B

Definition 5. a *relation* on A

Definition 8. a *function* f from X to Y

Definition 8. the *domain* of a function f

Definition 8. the *codomain* of a function f

Definition 8. the *graph* of a function f

Definition 11. the *image* of S under f

Definition 13. the *range* of a function f

Definition 18. functions f and g are *equal*

Definition 20. the *restriction* of f to S

Definition 25. f is *one-to-one* (or *injective*)

Definition 25. f is an *injection*

Definition 27. f is *onto* (or *surjective*)

Definition 27. f is a *surjection*

Definition 29. f is *bijective* (or a *one-to-one correspondence*)

Definition 29. f is a *bijection*

Definition 33. *composition* of a function f followed by a function g ($f(X) \subset W$)

Definition 33. the *domain* of $g \circ f$ ($f(X) \subset W$)

Definition 33. the *codomain* of $g \circ f$ ($f(X) \subset W$)

Definition 41. f is the *identity* function on X

Definition 43. f and g are *inverse* functions

Definition 43. f is an *invertible* function

Definition 48. the *inverse image* of T under f

Definition 54. *composition* of a relation f followed by a relation g

Definition 55. *composition* of a function f followed by a function g

Definition 55. the *domain* of $g \circ f$

Definition 55. the *codomain* of $g \circ f$

Supplemental Exercise 1. Find the range of:
$$f(x) = \frac{x^2}{x^2 + x - 2}$$
$$g(x) = |x^3 + x + 1|$$

$$h(x) = 2\sin\frac{x^2}{x^2+1}.$$

Supplemental Exercise 2. How many one-to-one functions can be defined from $\{1,2,3\}$ to $\{1,2,3,4\}$?

Supplemental Exercise 3. Determine which functions are 1-1:
$$f(x) = \frac{3x+1}{2x-3}$$
$$g(x) = \sqrt{1+\frac{1}{x}}$$
$$h(x) = |2x+1|$$
$$t(x) = \sqrt{-x^6}$$

Supplemental Exercise 4. Determine which functions are bijections from $\mathbb{R}$ to $\mathbb{R}$:
$$f(x) = ax + b \text{ where } a \neq 0$$
$$g(x) = |x|$$
$$h(x) = x^3 + 1$$

Supplemental Exercise 5. Let
$$f(x) = \frac{x+2a}{x-b} \quad \text{and} \quad g(x) = \frac{ax+2b}{x+b-2}.$$
Find all real values of the constants a and b such that $f = g$.

Supplemental Exercise 6. Let $S \subset \mathbb{R}$ and let $f : S \to \mathbb{R}$ be a function. Give a geometrical interpretation of what it means for f to be one-to-one.

Supplemental Exercise 7. Let $S \subset \mathbb{R}$ and let $f : S \to \mathbb{R}$ be a function. Give a geometrical interpretation of what it means for f to be onto.

Supplemental Exercise 8. Let $f : \mathbb{R} \to \mathbb{R}$ be a function with $f(x) = 2x + 1$. Prove that $f(\mathbb{N}) \subset f(\mathbb{R})$. Find a function $g : \mathbb{R} \to \mathbb{R}$ such that $g(\mathbb{N}) = g(\mathbb{R})$.

Supplemental Exercise 9. If $(f \circ g)(x) = x^2 + 3$ and $g(x) = 2x - 1$, find f.

Supplemental Exercise 10. If $(f \circ g)(x) = 3x + 2$ and $f(x) = 4x - 1$, find g.

Supplemental Exercise 11. Determine the values of a and b for which $f(x) = ax + b$ is invertible and find its inverse.

Supplemental Exercise 12. Determine the values of a, b, and c for which
$$f(x) = ax^2 + bx + c$$
is invertible and find its inverse.

Supplemental Exercise 13. Find a restriction $f|_S$ of $f(x) = 2x^2 - 5x + 1$ that is invertible. Prove or disprove that for a function f, there exists a restriction of f that is invertible.

Supplemental Exercise 14. Suppose that $f(x + y) = f(x) + f(y)$ for all real numbers x and y. Prove that
(a) $f(0) = 0$,

5.7. Supplemental Exercises

(b) $f(-x) = -f(x)$ for all $x \in \mathbb{R}$,

(c) $f(nx) = nf(x)$ for all $n \in \mathbb{N}$ and for all $x \in \mathbb{R}$.

Supplemental Exercise 15. Let $f(x) = ax + b$ and $g(x) = 2x - 3$. Determine a and b such that $f \circ g = g \circ f$.

Supplemental Exercise 16. Let $f(x) = 5x - 4$. Find $f^{-1}([2, 3])$ and $f^{-1}(\mathbb{R})$.

Supplemental Exercise 17. Let $f : \mathbb{R} - \{-2\} \to \mathbb{R}$ be a function, with

$$f(x) = \frac{2x - 7}{x + 2}.$$

Prove that f is not onto. Restrict the codomain so that f is onto. Prove that with the restricted codomain f is invertible and find its inverse.

Supplemental Exercise 18. Let f be a set of ordered pairs. Suppose that for all (x, y) and (x', y') in f, if $y \neq y'$, then $x \neq x'$. Let $X = \{x \mid (x, y) \in f\}$ and $Z = \{y \mid (x, y) \in f\}$. Prove that, for $Y \supset Z$, f is a function from X to Y.

Supplemental Exercise 19. Functions $f : X \to Y$ and $g : U \to V$ are equal if and only if $Y = V$ and $f = g$. That is, we can omit the condition that the domains are equal in Definition 18.

Supplemental Exercise 20. Let f be a function. Prove that f is one-to-one if and only if $f(A \cap B) = f(A) \cap f(B)$ for all subsets A and B of $\mathcal{D}_f$.

Supplemental Exercise 21. Find examples of three functions with codomain $\mathbb{R}$ such that the first is bijective, the second is one-to-one but not onto, and the third is neither one-to-one nor onto.

Supplemental Exercise 22. Suppose that $g \circ f$ is the composition of $f : X \to Y$ and $g : Y \to Z$. Suppose that $g \circ f$ is one-to-one.
(a) Prove that f is one-to-one.
(b) Find a counterexample to show that g need not be one-to-one.
(c) Prove that if f is onto, then f is bijective and g is one-to-one.

Supplemental Exercise 23. Suppose that $g \circ f$ is the composition of $f : X \to Y$ and $g : Y \to Z$. Suppose that $g \circ f$ is onto.
(a) Prove that g is onto.
(b) Find a counterexample to show that f need not be onto.
(c) Prove that if g is one-to-one, then g is bijective and f is onto.

Supplemental Exercise 24. Prove that for $f : X \to Y$ and $g : W \to X$, $f \circ I_X = f$ and $I_X \circ g = g$.

Supplemental Exercise 25. Let $f(x) = x - 6\sqrt{x} + 9$. Find the domain of f. Show that $(f \circ f)(x) = x$ for $x \in [0, 9]$. Is f invertible with $f^{-1} = f$?

Supplemental Exercise 26. Let $f : A \to A$ and $g : A \to A$ be invertible functions. Show that $(f \circ g)^{-1} = g^{-1} \circ f^{-1}$.

Supplemental Exercise 27. Let
$$f(x) = \begin{cases} x, & x \in \mathbb{Q} \\ x+1, & x \notin \mathbb{Q}. \end{cases}$$
Prove that f is invertible and find its inverse.

Supplemental Exercise 28. Let
$$f(x) = \begin{cases} x^2, & x \in \mathbb{Q} \\ x+1, & x \notin \mathbb{Q}. \end{cases}$$
Is f invertible? Explain your answer.

Supplemental Exercise 29. Suppose that $f(f(x)) = f(x) - 3x + 1$ for $x \in \mathbb{R}$. Prove that f is one-to-one and $f^{-1}(x) = \frac{1}{3}(x - f(x) + 1)$.

Supplemental Exercise 30. Suppose that
$$f(x) = \frac{x}{|x| + 3}$$
for $x \in \mathbb{R}$. Prove that f is one-to-one and find f^{-1}.

Supplemental Exercise 31. Let $f : X \to Y$ be a function.
(a) Prove that if $f^{-1}(f(S)) = S$ for every subset S of X, then f is one-to-one.
(b) Prove that if $f(f^{-1}(T)) = T$ for every subset T of Y, then f is onto.
(Compare with Supplemental Exercises 33 and 34.)

Supplemental Exercise 32. Let $f : X \to Y$ be a function with $A \subset Y$ and $B \subset Y$. Prove:
(a) $f^{-1}(A \cap B) = f^{-1}(A) \cap f^{-1}(B)$.
(b) $f^{-1}(A \cup B) = f^{-1}(A) \cup f^{-1}(B)$.
(c) $f^{-1}(A - B) = f^{-1}(A) - f^{-1}(B)$.

Supplemental Exercise 33. Let $f : X \to Y$ be a function with $S \subset X$.
(a) Prove that $f^{-1}(f(S)) \supset S$.
(b) Prove that if f is one-to-one, then $f^{-1}(f(S)) = S$.
(c) Is the converse of the statement in (b) true? Prove it or find a counterexample.

Supplemental Exercise 34. Let $f : X \to Y$ be a function with $T \subset Y$.
(a) Prove that $f(f^{-1}(T)) \subset T$.
(b) Prove that if f is onto, then $f(f^{-1}(T)) = T$.
(c) Is the converse of the statement in (b) true? Prove it or find a counterexample.

Supplemental Exercise 35. Let $f : X \to Y$ be a function. Prove that f is one-to-one if and only if for all $y \in f(X)$ there exists a unique $x \in X$ such that $f(x) = y$.

Supplemental Exercise 36. Project. Find earlier definitions of the concept of function by the mathematicians who first worked with this idea. Do they differ with the definition in this text? How?

6

Relations on a Set

6.1 Properties of Relations

In Chapter 5, we defined function as a kind of relation and *relation* from a set A to a set B as a subset $R \subset A \times B$. If $A = B$, we call R a relation on A. The examples of relations we looked at in Chapter 5 are relations on a single set. In fact, we are more interested in relations on a set and consider various properties that a relation may or may not have.

Definition 1. Let R be a relation on a set S.
(a) R is *reflexive* if and only if $s \, R \, s$ for all $s \in S$.
(b) R is *nonreflexive* if and only if $s \, \not{R} \, s$ for all $s \in S$.
(c) R is *symmetric* if and only if $s \, R \, t$ implies $t \, R \, s$ for all $s, t \in S$.
(d) R is *asymmetric* if and only if $s \, R \, t$ implies $t \, \not{R} \, s$ for all $s, t \in S$.
(e) R is *antisymmetric* if and only if $s \, R \, t$ and $t \, R \, s$ implies $s = t$ for all $s, t \in S$.
(f) R is *transitive* if and only if $s \, R \, t$ and $t \, R \, u$ implies $s \, R \, u$ for all $s, t, u \in S$.
(g) R is *connected* if and only if $s \, R \, t$ or $t \, R \, s$ or $s = t$ for all $s, t \in S$.
(h) R is *trichotomous* if and only if exactly one of $s \, R \, t, t \, R \, s, s = t$ is true for all $s, t \in S$.

Remark 2. Be careful with nonreflexive, asymmetric, and antisymmetric.

If R is a nonreflexive relation on a nonempty set, R is not reflexive. This is because R is not reflexive means $s \, \not{R} \, s$ for *some* s while R is nonreflexive means $s \, \not{R} \, s$ for *all* s.

For a relation R, neither R is asymmetric nor R is antisymmetric is equivalent to R is not symmetric. (See the Supplemental Exercises.)

In the definition of connected, we say that s and t are *comparable* if and only if $s \, R \, t$ or $t \, R \, s$ or $s = t$. There are many relationships between the properties in Definition 1; the most obvious is that if a relation is trichotomous, then it must be connected.

Example 3. Let R be the relation given by $x^2 + y^2 = 1$ in Example 6 of Chapter 5. Then:
- R is not reflexive since 1 is not related to 1: $1^2 + 1^2 = 2 \neq 1$.
- R is not nonreflexive since $1/\sqrt{2}$ is related to $1/\sqrt{2}$: $(1/\sqrt{2})^2 + (1/\sqrt{2})^2 = 1$. (Not nonreflexive does not imply reflexive.)
- R is symmetric by the commutative property of addition.
- R is not asymmetric (see Supplemental Exercise 1).
- R is not antisymmetric since $0^2 + 1^2 = 1^2 + 0^2 = 1$, but $0 \neq 1$.

- R is not transitive since 1 is related to 0 and 0 is related to 1, but 1 is not related to 1.
- R is neither connected nor trichotomous since 1 and $\frac{1}{2}$ satisfy none of the conditions.

A relation on the empty set satisfies all of the properties (vacuously) of Definition 1. On a nonempty set, some of these properties conflict with or guarantee some other property such as in the following:

Proposition 4. *A relation on a nonempty set cannot be both reflexive and nonreflexive.*

Proof. Let R be a relation on a nonempty set S. Suppose that $x \in S$. Assume that R is reflexive and nonreflexive. Since R is reflexive, $x\ R\ x$. Since R is nonreflexive, $x\ \not R\ x$. That is, $(x, x) \in R$ and $(x, x) \notin R$. This is a contradiction. □

For the remainder of this chapter, we concentrate on two types of relations on a set. First, we will consider order relations, which generalize "less than" and "less than or equal to." Second, we will consider equivalence relations, which generalizes "is equal to."

Before we generalize, let's review what we know about $<$, $\leq$, and $=$ on $\mathbb{R}$ and $\subsetneq$ and $\subset$ on $\mathcal{P}(U)$, the power set of a nonempty set U. We will see that $<$ and $\subsetneq$ are of the same type, as are $\leq$ and $\subset$.

> **Exercise 5.** Determine whether the relation $<$ on $\mathbb{R}$ satisfies or fails to satisfy each of the eight properties of relations given in Definition 1. Prove your answers.

The relation $\subsetneq$ on $\mathcal{P}(U)$ behaves similarly to the relation $<$ on $\mathbb{R}$. In the answer to Exercise 5, substituting $\subsetneq$ and $\mathcal{P}(U)$ for $<$ and $\mathbb{R}$, respectively, give proofs concerning the properties of $\subsetneq$.

> **Exercise 6.** Determine whether the relation $\subset$ on $\mathcal{P}(U)$ for some nonempty U satisfies or fails to satisfy each of the eight properties of relations given in Definition 1. Prove your answers.

Again, the answer to Exercise 6 can be imitated for $\leq$ on $\mathbb{R}$.
Next, we examine equality on $\mathbb{R}$ (or any nonempty set).

> **Exercise 7.** Determine whether the relation $=$ on $\mathbb{R}$ satisfies or fails to satisfy the eight properties of relations given in Definition 1. Prove your answers.

One interesting feature of equality is that it is both symmetric and antisymmetric. Keep the difference between antisymmetric and asymmetric.

6.2 Order Relations

A natural kind of relation is one that puts an order on the set. The relations $<$ on $\mathbb{R}$ and $\subset$ on $\mathcal{P}(U)$ (which were considered in Exercises 5 and 6) are examples. We start with two definitions that model these two examples, but are more general.

6.2. Order Relations

Definition 8. A relation $\prec$ on a set S that is nonreflexive and transitive is called a *strict partial ordering*. A *strictly partially ordered* set is a set S together with a strict partial ordering $\prec$ on S.

Definition 9. A relation $\preceq$ on a set S that is reflexive, antisymmetric, and transitive is called a *partial ordering*. A *partially ordered* set is a set S together with a partial ordering $\preceq$ on S.

Using bold characters, a **partially ordered set** is sometimes called a *poset*.

Exercise 10. Does $=$ as a relation on $\mathbb{R}$ define a strict partial ordering? Does it define a partial ordering? Prove your answers.

The example in Exercise 10 is rather trivial since most elements are not comparable.

Exercise 11. Consider $\mathbb{R}$ with $<$ and $\mathcal{P}(U)$ with $\subset$. Which is a strictly partially ordered set? Which is a partially ordered set? Explain your answers.

Sometimes, especially for finite partially ordered sets in which not all elements are comparable, it is useful to make an illustration, called a Hasse diagram (Figure 6.1), of the set with its relation.

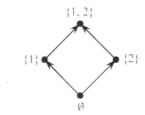

Figure 6.1. The Hasse diagram for $\mathcal{P}(\{1,2\})$ with $\subset$

Figure 6.1 shows the elements of $\mathcal{P}(\{1,2\})$ with the relation $\subset$. Elements of $\mathcal{P}(\{1,2\})$ are shown as the vertices of a graph. Arrows show the order relation; if there is a directed path from one vertex to another using one or more arrows, then the vertex at the tail of the (first) arrow is related to the vertex at the head of the (last) arrow. For example, the arrow from $\emptyset$ to $\{1\}$ indicates that $\emptyset \subset \{1\}$ and the arrow from $\emptyset$ to $\{2\}$ together with the arrow from $\{2\}$ to $\{1,2\}$ indicate that $\emptyset \subset \{1,2\}$.

It is possible to give a precise definition of a Hasse diagram in terms of directed graphs, but we will not do that.

Using the convention that arrows always go up simplifies the presentation. Figure 6.2 shows the more complicated Hasse diagram for $\mathcal{P}(\{1,2,3\})$; you should see it as a cube.

The next exercise shows that strict partial orderings and partial orderings are intimately related.

Exercise 12. (a) Suppose that $\preceq$ is a partial ordering on S and define the relation $\prec$ on S by $s \prec t$ if and only if $s \preceq t$ and $s \neq t$. Show that $\prec$ is a strict partial ordering on S.
(b) Suppose that $\prec$ is a strict partial ordering on S and define the relation $\preceq$ on S by $s \preceq t$ if and only if $s \prec t$ or $s = t$. Show that $\preceq$ is a partial ordering on S.

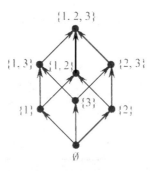

Figure 6.2. The Hasse diagram for $\mathcal{P}(\{1,2,3\})$ with $\subset$

Remark 13. According to Exercise 12, it does not matter whether the order on a partially ordered set S is derived from a partial ordering $\preceq$ or a strict partial ordering $\prec$ on S. Put another way, whenever either type of order relation is present, both are. Hence, we can use whichever is more convenient. Some authors may be sloppy and write "a partial ordering $\prec$" when "a strict partial ordering $\prec$" is meant.

Definition 14. A strict partial ordering $\prec$ on S that is connected is called a *strict total ordering*. A *strictly totally ordered* set is a set S together with a strict total ordering $\prec$ on S.

Definition 15. A partial ordering $\preceq$ on S that is connected is called a *total ordering*. A *totally ordered* set is a set S together with a total ordering $\preceq$ on S.

> **Exercise 16.** (a) Prove that every strict total ordering $\prec$ on a set S is trichotomous.
> (b) Suppose that R is a relation on S that is transitive and trichotomous. Prove that R is a strict total ordering.

That is, a relation is a strict total ordering if and only if it is transitive and trichotomous.

In our notation, we will always assume that we can go back and forth between $\prec$ or $\preceq$, as in Exercise 12. Can we use $\succ$ or $\succeq$ when we have $\prec$ or $\preceq$, respectively? The answer is yes, of course. That is, $x \succ y$ is defined as $y \prec x$ and $x \succeq y$ is defined as $y \preceq x$.

The following definitions are related, but are not the same.

Definition 17. Let S be a partially ordered set with partial ordering $\preceq$. An element $m \in S$ is *maximal* if and only if there exists no $s \in S$ such that $m \prec s$.

Definition 18. Let S be a partially ordered set with partial ordering $\preceq$. An element $M \in S$ is *greatest* if and only if $s \preceq M$ for all $s \in S$.

Definitions 17 and 18 are different. In a partially ordered set, elements need not be related. The following example shows the difference between maximal and greatest elements.

Example 19. Let $A = \{1, 2\}$. Let $S = \mathcal{P}(A)$ and $T = S - \{A\}$ with the partial ordering $\subset$. S has A as its only maximal element and its only greatest element. T has two maximal elements, $\{\{1\}\}$ and $\{\{2\}\}$, but no greatest element.

Example 20. $\mathbb{N}$ with the partial ordering $\leq$ has neither a maximal element nor a greatest element.

6.2. Order Relations

Maximal and greatest elements are intimately related as we will see from the next three exercises. First, we show that a partially ordered set can have only one greatest element.

Exercise 21. Let S be a partially ordered set with partial ordering $\preceq$. Prove that if S has a greatest element, then it is unique.

Second, we show that a greatest element of a partially ordered set is also a maximal element.

Exercise 22. Let S be a partially ordered set with partial ordering $\preceq$. Prove that the greatest element of S, if it exists, is a maximal element.

Third, we show that a maximal element of a totally ordered set is also a greatest element.

Exercise 23. Let S be a totally ordered set with total ordering $\preceq$. Prove that a maximal element of S, if it exists, is the greatest element.

In a totally ordered set, by Exercises 22 and 23, maximal and greatest elements are identical and, by Exercise 21, if a maximal element exists, then it is unique.

We can define minimal elements and the least element of a partially ordered set in analogy with the definitions of maximal and greatest elements. Analogous results can be proved.

Next we consider subsets in a partially ordered set. The partial ordering on the ambient set induces a partial ordering on its subsets.

Definition 24. Let A be a subset of S, where S is a partially ordered set with partial ordering $\preceq$. An element $m \in S$ is an *upper bound* on A if and only if $a \preceq m$ for all $a \in A$. A is *bounded above* if and only if A has an upper bound.

Definition 25. Let A be a subset of S, where S is a partially ordered set with partial ordering $\preceq$. An element $M \in S$ is a *least upper bound* on A if and only if M is an upper bound on A and $M \preceq s$ for all upper bounds s on A.

A least upper bound is also called a *supremum*. In Definitions 24 and 25, M is not necessarily an element of A.

Example 26. Let $A = \{x \mid 0 < x^2 < 2\}$. 2 is an upper bound on A. What is the least upper bound of A? We have not been precise enough to answer this. If we consider A as a subset of $\mathbb{R}$, then $\sqrt{2}$ is the least upper bound. If we consider A as a subset of $\mathbb{Q}$, there is no least upper bound! The decreasing sequence of rational upper bounds

$$2, 1.5, 1.42, 1.415, 1.4143, 1.41422, 1.414214, \ldots$$

converges to the irrational number $\sqrt{2}$ in $\mathbb{R}$. Since the limit is unique, there is no rational least upper bound.

Example 27. $\mathbb{N}$ with the partial ordering $\leq$ has no upper bounds. The set E of even natural numbers has no upper bounds.

> **Exercise 28.** Let S be a partially ordered set with partial ordering $\preceq$ and $A \subset S$. Prove that if A has a least upper bound, then it is unique.

We can define lower bounds and the greatest lower bound (or infimum) of a subset of a partially ordered set in analogy with the definitions of upper bound and least upper bound. Analogous results can be proved.

Definition 29. Let S be a partially ordered set with partial ordering $\preceq$. S has the *least upper bound property* if and only if every nonempty subset that is bounded above has a least upper bound.

The least upper bound property is important. It distinguishes the rational numbers from the real numbers and will be discussed more in Chapters 8 and 12.

Example 30. If $A = \{x \in \mathbb{Q} \mid 0 < x^2 < 2\}$, then $\sqrt{2}$ is the least upper bound of A as a subset of $\mathbb{R}$ and A has no least upper bound as a subset of $\mathbb{Q}$. Hence $\mathbb{Q}$ does not have the least upper bound property.

We mention here one more definition related to order.

Definition 31. A totally ordered set S is *well-ordered* if and only if every nonempty subset of S has a least element. The ordering is called a *well-ordering*.

The fact that $\mathbb{N}$ is well-ordered is proved in Exercise 8 of Chapter 9. The question of how to well-order a set in general is difficult. The statement that every set has a well-ordering is equivalent to the Axiom of Choice, which we will see again in Chapter 7.

6.3 Equivalence Relations

The classification of the forms of life—first effected systematically by Aristotle—provides an example of grouping entities by shared characteristics. Early Euclidean geometry provides examples.

Example 32. The concept of triangle similarity divides the set of all triangles into mutually disjoint classes of triangles. We can see that

Every triangle is similar to itself.
If a triangle ABC is similar to a triangle DEF, then DEF is similar to ABC.
If a triangle ABC is similar to a triangle DEF and DEF is similar to the triangle GHI, then ABC is similar to GHI.

Rephrased in the language of this chapter, triangle similarity is a relation on the set of triangles that is reflexive, symmetric, and transitive. Relations of this kind are called equivalence relations.

This leads us to the definition.

Definition 33. A relation $\approx$ on a set S that is reflexive, symmetric, and transitive is called an *equivalence* relation.

An equivalence relation can be thought of as a generalization of the equality relation. Another interesting example of an equivalence relation follows.

6.3. Equivalence Relations

Example 34. Let $k \in \mathbb{N}$. Define the relation $\equiv$ on $\mathbb{N}$ by $n \equiv m$ if and only if $\frac{n-m}{k} \in \mathbb{Z}$. We usually say that *n is congruent to m modulo k* and write

$$n \equiv m \pmod{k}.$$

This equivalence relation is well known to children, who learned it in the form that 3 hours after 11 o'clock comes 2 o'clock; in this case $k = 12$ and the statement is that

$$11 + 3 \equiv 2 \pmod{12}.$$

In Chapter 7, given two sets A and B, we will say that A has the same cardinality as B if and only if there exists a bijection $f : A \to B$.

Exercise 35. Prove that "has the same cardinality as" is an equivalence relation on the power set of a set.

Partitioning the set of clothes in your house into categories such as coats, shirts, pants, underwear, socks, and shoes is an example of an equivalence relation at work. Rather than giving the relation in terms of pairs of objects, it is given in terms of the subsets that supply the entries of the pairs. This leads us to the following definition.

Definition 36. Let $\approx$ be an equivalence relation on S and let $x \in S$. The set

$$[x]_\approx = \{s \in S \mid x \approx s\}$$

is called the *equivalence class* of x under $\approx$.

For an example, see Example 32.

Example 37. Under the equivalence relation of similarity on the set of triangles, the equivalence class of a triangle T is the set of all triangles similar to T.

The equivalence class of x is sometimes written as $[x]$ when the equivalence relation is understood.

Example 38. Under the equivalence relation of congruence modulo 12 on the integers, the equivalence class of 0 is

$$[0] = \{\ldots, -24, -12, 0, 12, 24, \ldots\}$$

and the equivalence class of 5 is

$$[5] = \{\cdots, -19, -7, 5, 17, 29, \ldots\}.$$

Take 0:00 as midnight, the beginning of the day. Then noon of the day and midnight (ending the day and beginning the next day) are both congruent modulo 12. That is noon and midnight can both be denoted 0:00 or 12:00. Also, 17:00 is the same as 5:00 (in the afternoon).

Next, we examine the nature of equivalence classes under an equivalence relation.

Proposition 39. *If $\approx$ is an equivalence relation on a set S, then $s \in [s]_\approx$ for every $s \in S$.*

Proof. Choose $s \in S$. Since $\approx$ is an equivalence relation, $s \approx s$ by reflexiveness. By the definition of equivalence class, $s \in [s]_{\approx}$. □

Proposition 39 is useful. We use it to prove the following theorem.

Theorem 40. *Let $\approx$ be an equivalence relation on a set S and let $s, t \in S$. Then $[s] = [t]$ if and only if $s \approx t$.*

Proof. Suppose that $[s] = [t]$. By Proposition 39, $t \in [s]$. By definition of equivalence relation, $s \approx t$.

Conversely, suppose that $s \approx t$. To show that $[s] = [t]$, we first show that $[t] \subset [s]$. Suppose that $x \in [t]$. Hence, $t \approx x$. By transitivity, $s \approx x$. Thus, $x \in [s]$. Therefore, $[t] \subset [s]$.

The verification of $[s] \subset [t]$ is left as Exercise 41. □

Exercise 41. Use the symmetry of $\approx$ to complete the proof of Theorem 40.

What happens if $s \not\approx t$?

Exercise 42. Let $\approx$ be an equivalence relation on a set S and let $s, t \in S$. Prove that if $[s] \neq [t]$, then $[s] \cap [t] = \emptyset$, that is, $[s]$ and $[t]$ are disjoint.

The converse of Exercise 42 is addressed in Supplemental Exercise 14. Equivalence classes of the same equivalence relation are nonempty and pairwise disjoint. We will rephrase these results using the following language.

Definition 43. Let S be a nonempty set. A *partition* of S is a set of nonempty, pairwise disjoint subsets of S whose union is S.

Example 44. Let us give three partitions of $\mathbb{Z}$, the integers.

The even integers and the odd integers partition $\mathbb{Z}$.
The natural numbers, the negative integers, and the singleton set $\{0\}$ partition $\mathbb{Z}$.
The singleton sets $\{n\}$ for all $n \in \mathbb{Z}$ partition $\mathbb{Z}$.

We next show that it is possible to get a partition from an equivalence relation.

Definition 45. Let $\approx$ be an equivalence relation on a nonempty set S. Denote by $S/\approx$ the *set of all equivalence classes of S under $\approx$*.

If we read $\approx$ as "equivalence," we read $S/\approx$ as "S mod equivalence." After the following exercise, we call $S/\approx$ *the partition of S induced by* $\approx$.

Exercise 46. Prove that if $\approx$ is an equivalence relation on a nonempty set S, then $S/\approx$ is a partition of S.

Next are two examples of partitions derived from equivalence relations.

Example 47. For the equivalence relation $=$ on a nonempty set S, $S/=$ is the partition of S into singleton sets.

6.3. Equivalence Relations

Example 48. For congruence modulo 3 on $\mathbb{Z}$,

$$\mathbb{Z}/\equiv (\bmod 3) = \{\,[0], [1], [2]\,\},$$

where $\mathbb{Z}/\equiv (\bmod 3)$ is the partition S mod the equivalence relation congruence modulo 12.

We have gone from equivalence relations to partitions. Next we go in the other direction.

Definition 49. Let $\mathcal{C}$ be a partition of a nonempty set S. Define a relation R on S by $s\,R\,t$ if and only if there exists $C \in \mathcal{C}$ such that $s \in C$ and $t \in C$. R is called *the equivalence relation induced by the partition* $\mathcal{C}$ and is denoted by $S/\mathcal{C}$.

It must be verified that $S/\mathcal{C}$ actually is an equivalence relation. We read $S/\mathcal{C}$ as "S mod $\mathcal{C}$" and $s\,S/\mathcal{C}\,t$ as "s is S mod $\mathcal{C}$-related to t." The verification is Exercise 50.

> **Exercise 50.** Let $\mathcal{C}$ be a partition of a nonempty set S. Prove that $S/\mathcal{C}$ is an equivalence relation on S.

Since we can pass from an equivalence relation to a partition and then back to an equivalence relation, we ask how the original relation and the induced relation are related. They are the same.

Theorem 51. *If $\approx$ is an equivalence relation on a nonempty set S, then*

$$S/(S/\approx) = \approx.$$

Proof. By Exercise 46, $S/\approx$ is a partition of S and, by Exercise 50, $S/(S/\approx)$ is an equivalence relation on S. Since a relation on S is a subset of $S \times S$, it suffices to show that $S/(S/\approx) \subset \approx$ and $\approx \subset S/(S/\approx)$.

Suppose that $(s, t) \in S/(S/\approx)$; that is, $s\,S/(S/\approx)\,t$. By Definition 49, there exists $C \in S/\approx$ such that $s, t \in C$. By Definition 45, there exists $u \in S$ such that $C = [u]$. Since $s, t \in [u]$, $u \approx s$ and $u \approx t$. Hence, $s \approx t$. Therefore, $S/(S/\approx) \subset \approx$.

The remainder of the proof is left for Supplemental Exercise 15. $\square$

We can do this starting with a partition as well.

Theorem 52. *If $\mathcal{C}$ is a partition of a nonempty set S, then*

$$S/(S/\mathcal{C}) = \mathcal{C}.$$

Proof. By Exercise 50, $S/\mathcal{C}$ is an equivalence relation on S, and, by Exercise 46, $S/(S/\mathcal{C})$ is a partition of S. Since a partition of S is a subset of $\mathcal{P}(S)$, it suffices to show that $S/(S/\mathcal{C}) \subset \mathcal{C}$ and $\mathcal{C} \subset S/(S/\mathcal{C})$.

Suppose $C \in \mathcal{C}$. Choose an element $s \in C$. Since $S/(S/\mathcal{C})$ is a partition of S, there exists $C' \in S/(S/\mathcal{C})$ such that $s \in C'$. $C' = [s]$ with respect to the equivalence relation $S/\mathcal{C}$. We must show that $C = [s]$.

First, fix $t \in C$. By Definition 49, $s\,S/\mathcal{C}\,t$. By Theorem 40, $t \in [t] = [s]$. So $C \subset [s]$.

Next, fix $t \in [s]$. So $s\, S/\mathcal{C}\, t$. By Definition 49, there exists $C'' \in S/(S/\mathcal{C})$ such that $s, t \in C''$. Since $S/(S/\mathcal{C})$ is a partition and $s \in C\hat{C}''$, $C'' = C$. So $t \in C$ and $[s] \subset C$. Hence, $C \in S/(S/\mathcal{C})$. Therefore, $\mathcal{C} \subset S/(S/\mathcal{C})$.

The remainder of the proof is left for Supplemental Exercise 16. □

6.4 Supplemental Exercises

Definition Review. Twenty-nine italicized terms were defined in this chapter.. Let S be a set with subset A and let R, $\prec$, $\preceq$, and $\approx$ be relations on S. Define each of the following and give an example of each:

Definition 1. R is *reflexive*

Definition 1. R is *nonreflexive*

Definition 1. R is *symmetric*

Definition 1. R is *asymmetric*

Definition 1. R is *antisymmetric*

Definition 1. R is *transitive*

Definition 1. R is *connected*

Definition 1. R is *trichotomous*

Definition 8. $\prec$ is a *strict partial ordering*

Definition 8. S is a *strictly partially ordered* set

Definition 9. $\preceq$ is a *partial ordering*

Definition 9. S is a *partially ordered* set

Definition 14. $\prec$ is a *strict total ordering*

Definition 14. S is a *strictly totally ordered* set

Definition 15. $\preceq$ is a *total ordering*

Definition 15. S is a *totally ordered* set

Definition 17. $m \in S$ is a *maximal* element

Definition 18. $m \in S$ is a *greatest* element

For 24–29, let S be a partially ordered set with partial ordering $\preceq$.

Definition 24. $m \in S$ is an *upper bound* on A

Definition 24. A is *bounded above*

Definition 25. $m \in S$ is a *least upper bound* on A

Definition 29. S has the *least upper bound property*

For 31, let S be a totally ordered set with total ordering $\preceq$.

Definition 31. S is *well-ordered*

Definition 31. $\preceq$ is a *well-ordering*

Definition 33. $\approx$ is an *equivalence* relation

Definition 36. the *equivalence class of x under $\approx$*, an equivalence relation

Definition 43. a *partition* of a nonempty set, S

For 45, let $\approx$ be an equivalence relation on S.

Definition 45. $S/\approx$, the *set of all equivalence classes of S under $\approx$*

For 49, let $\mathcal{C}$ be a partition of S.

Definition 49. $S/\mathcal{C}$, the equivalence relation *induced* by $\mathcal{C}$

Supplemental Exercise 1. Prove that a relation on a nonempty set cannot be both symmetric and asymmetric.

Supplemental Exercise 2. Give an example of a relation that is asymmetric but not symmetric.

Supplemental Exercise 3. Prove that a relation on a set that is trichotomous is also connected.

Supplemental Exercise 4. Give an example of a relation on a nonempty set that is both symmetric and antisymmetric.

Supplemental Exercise 5. Let $R = \{(1,1)\}$ be a relation on $\{1\}$. Which of the properties given in Definition 1 hold and which do not? Prove your answers.

Supplemental Exercise 6. Let $R = \{(1,1),(2,3)\}$ be a relation on $\{1,2,3\}$. Which of the properties given in Definition 1 hold and which do not? Prove your answers.

Supplemental Exercise 7. Let $R = \{(1,1),(2,3)\}$ be a relation on $\{1,2,3,4\}$. Which of the properties given in Definition 1 hold and which do not? Prove your answers.

Supplemental Exercise 8. Let X be the set of all 2×2 matrices with real entries. The determinant of a matrix $\begin{pmatrix} a & b \\ c & d \end{pmatrix}$ equals $ad - bc$. Denote the determinant of a matrix M by $|M|$. Define a relation $<$ on X by $A < B$ if and only if $|A| < |B|$. Prove that $<$ is a strict partial ordering on X. Is $<$ a total ordering? Explain your answer.

Supplemental Exercise 9. Give examples of relations that have any combination of the properties given in Definition 1.

Supplemental Exercise 10. Let S be a partially ordered set with partial ordering $\preceq$. Prove that if S has more than one maximal element, then S does not have a greatest element.

Supplemental Exercise 11. Give an example of a nonempty subset A of a partially ordered set S such that the least upper bound of A exists and equals the greatest element of A. Prove your answer.

Supplemental Exercise 12. Give an example of a nonempty subset A of a partially ordered set S such that the least upper bound of A exists but is not the greatest element of A. Prove your answer.

Supplemental Exercise 13. Let A be a subset of a partially ordered set S with partial ordering $\preceq$. Prove that if A has a greatest element, then the least upper bound of A equals the greatest element of A.

Supplemental Exercise 14. Prove the converse of Exercise 42.

Supplemental Exercise 15. Complete the proof of Theorem 51: prove that $\approx \, \subset S/(S/\approx)$.

6.4. Supplemental Exercises

Supplemental Exercise 16. Complete the proof of Theorem 52: prove that $S/(S/\mathcal{E}) \subset \mathcal{E}$.

Supplemental Exercise 17. Let R be a relation on $\mathbb{N}$ defined by $p \; R \; q$ if and only if $p + q$ is prime. Does R satisfy or fail to satisfy the eight properties of relations given in Definition 1? Prove your answers.

Supplemental Exercise 18. Let R be a relation on $\mathbb{R}^2$, defined by $x \; R \; y$ if and only if
$$(x + y - 1)^2 - (2x - y)^2 = 0.$$
Does R satisfy or fail to satisfy the eight properties of relations given in Definition 1? Prove your answers.

Supplemental Exercise 19. Let R be a relation on $\mathbb{R}^2$, defined by $x \; R \; y$ if and only if
$$(x + y - 1)^2 + (2x - y)^2 = 0.$$
Does R satisfy or fail to satisfy the eight properties of relations given in Definition 1? Prove your answers.

Supplemental Exercise 20. Let $S = \{f : \mathbb{R} \to \mathbb{R} \mid f \text{ is bijective}\}$. Define a relation $*$ on S, by $f * g$ if and only if $(f \circ g)(x) = x$, for all $x \in \mathbb{R}$. Does $*$ satisfy or fail to satisfy the eight properties of relations given in Definition 1? Prove your answers.

Supplemental Exercise 21. Prove that the relations are equivalence relations:
  $x \; R \; y$ if and only if $|x - y| < 1$, for all real numbers x and y.
  $x \; R \; y$ if and only if $\sin x = \sin y$, for all real numbers x and y.
  $x \; R \; y$ if and only if there exists an integer k such that $x = y + k$, for all real numbers x and y.
  $x \; R \; y$ if and only if $x - y$ is even, for all integers x and y.
In the first three cases find the equivalence class of the number π.

Supplemental Exercise 22. Can you define an equivalence relation that partitions R into two classes? Explain.

Supplemental Exercise 23. Let $S = \{n \in \mathbb{N} \mid |n - 2| < 3\}$. Investigate whether it is possible to define an equivalence relation on S.

Supplemental Exercise 24. From the partition of $\mathbb{N}$
$$\{\{1\},\{2,3\},\{4,5,6\},\{7,8,9,10\},\dots\},$$
define the equivalence relation induced by this partition.

Supplemental Exercise 25. Let $S = \{1, 2, 3, 4, 5\}$. Define a relation
$$R = \{(x, y) \in S \times S \mid x + y = 4\}.$$
Are there maximal elements in S? Is there a greatest element?

Supplemental Exercise 26. Let D_f and D_g be the domains of the functions
$$f(x) = \frac{1}{1 + \sqrt{4 - |x|}} \quad \text{and} \quad g(x) = \frac{x^{2011}}{\sqrt{2 - x^2}},$$

respectively. Examine $\mathcal{D}_f \cap \mathcal{D}_g$ for maximal elements, greatest elements, least upper bounds, and greatest lower bounds, with the usual partial ordering $\leq$.

Supplemental Exercise 27. Let

$$S = \left\{ (-1)^n - \frac{1}{n} \,\middle|\, n \in \mathbb{N} \right\}$$

be the partially ordered set with the usual partial ordering $\leq$. Examine S for maximal elements, greatest elements, least upper bounds, and greatest lower bounds.

Supplemental Exercise 28. Project. Investigate how the logician and mathematical philosopher Bertrand Russell defined natural number in terms of equivalence relations.

7

Cardinality

7.1 Cardinality of Sets: Introduction

It seems reasonable to say that the two sets

$$A = \{a, b, c\} \quad \text{and} \quad B = \{\delta, \varepsilon, \zeta\},$$

have the same "size." Perhaps we should say what "size" means exactly. Both sets have exactly three elements. The idea of counting is an advanced concept compared to what is needed here. A more primitive idea would be to compare the sets by matching their

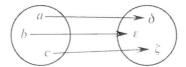

Figure 7.1. A one-to-one correspondence

elements as in the following:

$$a \leftrightarrow \delta \quad b \leftrightarrow \varepsilon \quad c \leftrightarrow \zeta.$$

This produces a one-to-one correspondence, or bijective function, between the sets. This is a fancy description for a simple idea.

The comparison of infinite sets such as $\mathbb{N}$, $\mathbb{Z}$, $\mathbb{Q}$, $\mathbb{R}$, and $\mathbb{C}$ is trickier. Many people have been intimidated by the notion of infinity. The ideas of one-to-one correspondence and counting, which we considered for finite sets, lead to analogous ideas about infinite sets. In the infinite case, the notion of one-to-one correspondence is no more difficult, while the notion of "infinite numbers" to represent the "size" of infinite sets is more difficult.

To see why people have been intimidated by infinity, consider one of the paradoxes of Zeno, an ancient Greek philosopher. According to Zeno, you can never reach the end of a racecourse since you must first cover the first half before you can cover the whole course, and half of the remainder before covering the second half of the course, and so on. That is, you cannot start from 0, go through $\frac{1}{2}$, $\frac{3}{4}$, $\frac{7}{8}$, $\frac{15}{16}$, etc., and arrive at 1 in finite time. This paradox has been around for nearly 2500 years. The paradox—and the broader concept of infinity—points to the foundations of mathematics and issues unresolved until the end of the nineteenth century.

7.2 Finite Sets

We would like to formalize the definition of the "size" of a set, but we do something a little different.

Definition 1. Sets A and B *have the same cardinality* if and only if there exists a one-to-one correspondence $f : A \to B$.

The sets $A = \{a, b, c\}$ and $B = \{\delta, \varepsilon, \zeta\}$ have the same cardinality since $f : A \to B$ defined by
$$f(a) = \delta \qquad f(b) = \varepsilon \qquad f(c) = \zeta$$
is a one-to-one correspondence. Notice that we did not say what "cardinality" is, only what "same cardinality" is. That is, we are saying that A and B have the same number of elements without saying what that number is!

Example 2. Infinite sets are more interesting. Let
$$\mathbb{N} = \{1, 2, 3, 4, \ldots\} \qquad \text{and} \qquad E = \{2, 4, 6, 8, \ldots\}.$$

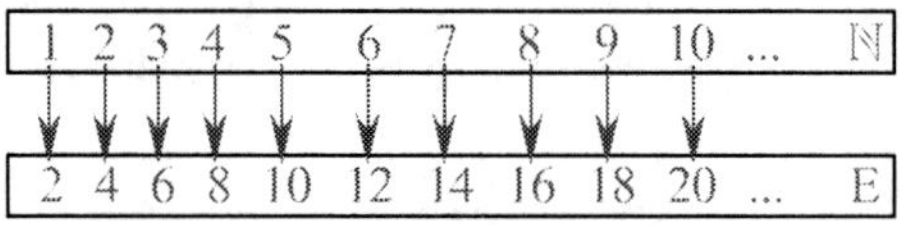

Figure 7.2. A one-to-one correspondence from $\mathbb{N}$ to E

Then $E \subsetneq \mathbb{N}$. If you think that E and $\mathbb{N}$ are not the same "size," then you are not thinking of the cardinality of the sets! It is easy to see that $f : \mathbb{N} \to E$ given by $f(n) = 2n$ is a one-to-one correspondence. Hence E and $\mathbb{N}$ have the same cardinality.

Let us turn our attention to finite sets.

Exercise 3. Let A be a set. Prove that A and $\emptyset$ have the same cardinality if and only if $A = \emptyset$.

The next three exercises are important consequences of facts about bijective functions.

Exercise 4. Let A be a set. Prove that A and A have the same cardinality.

Exercise 5. Let A and B be sets. Prove that A and B have the same cardinality if and only if B and A have the same cardinality.

Exercise 6. Let A, B, and C be sets. Prove that if A and B have the same cardinality and B and C have the same cardinality, then A and C have the same cardinality.

The properties of "have the same cardinality" in Examples 4, 5, and 6 are called, respectively, reflexivity, symmetry, and transitivity. These are the properties of equivalence

7.2. Finite Sets

relations, which we studied in Section 6.3 of Chapter 6. That is, "have the same cardinality" is an equivalence relation on a set of sets.

For $n \in \mathbb{N}$, let $\mathbb{Z}_n = \{1, 2, 3, \ldots, n\}$. We will use $\mathbb{Z}_n$ to further our understanding of finite sets. For example, if A and $\mathbb{Z}_n$ have the same cardinality, then we can think of $\mathbb{Z}_n$ as an index for A. So A looks like $\{a_1, a_2, a_3, \ldots, a_n\}$.

How do we know if

$$\{a_1, a_2, a_3, \ldots, a_m\} \quad \text{and} \quad \{b_1, b_2, b_3, \ldots, b_n\}$$

have the same cardinality? The answer is whenever $m = n$. This can be proved by taking one element at a time away from each set until one set is empty. Then you check if the other is also empty. We will turn this into an induction argument. We do this first for the a special case.

Theorem 7. $\mathbb{Z}_m$ and $\mathbb{Z}_n$ have the same cardinality if and only if $m = n$.

Proof. That $m = n$ implies $\mathbb{Z}_m$ and $\mathbb{Z}_n$ have the same cardinality follows from the fact that the identity function on $\mathbb{Z}_m$ is a bijection.

The proof in the other direction is by induction on n; that is, n is fixed in both steps of the proof while m is variable.

Consider the step $n = 1$. Then there exists a one-to-one correspondence $f : \mathbb{Z}_1 \to \mathbb{Z}_m$. Assume $m \neq 1$. Hence, $m \geq 2$ and $1, 2 \in \mathbb{Z}_m$. Since f is onto, $f(1) = 1$ and $f(1) = 2$. This contradiction of the one-to-one property shows that $m = n$.

To prove the inductive step, we suppose the induction hypothesis: $\mathbb{Z}_m$ and $\mathbb{Z}_n$ have the same cardinality implies $m = n$. Suppose that $\mathbb{Z}_m$ and $\mathbb{Z}_{n+1}$ have the same cardinality. Thus, there exists a one-to-one correspondence $f : \mathbb{Z}_{n+1} \to \mathbb{Z}_m$. Since $1 \neq n+1$, $f(1)$ and $f(n+1)$ are distinct elements of $\mathbb{Z}_m$ and $m - 1 \in \mathbb{N}$. Let $g : \mathbb{Z}_m \to \mathbb{Z}_m$ be defined by

$$g(x) = \begin{cases} f(n+1), & \text{if } x = m \\ m, & \text{if } x = f(n+1) \\ x, & \text{otherwise.} \end{cases}$$

(If $f(n+1) = m$, then $g = I_{\mathbb{Z}_m}$.) Since f and g are bijections, $g \circ f : \mathbb{Z}_{n+1} \to \mathbb{Z}_m$ is a one-to-one correspondence. Since

$$g(f(n+1)) = m,$$

we can define $h : \mathbb{Z}_n \to \mathbb{Z}_{m-1}$ by $h(x) = g(f(x))$. This is a one-to-one correspondence. By the induction hypothesis, $m - 1 = n$. Hence, $m = n + 1$. $\square$

We are now in a position to define what a finite set is.

Definition 8. A set S is *finite* if and only if either $S = \emptyset$ or there exists some $n \in \mathbb{N}$ such that S and $\mathbb{Z}_n$ have the same cardinality. We say that S *has cardinality* 0 *or* n, respectively, and denote this by $|S| = 0$ or $|S| = n$.

Exercise 9. Suppose A and B are finite sets with cardinalities n and m, respectively. Prove that A and B have the same cardinality if and only if $n = m$.

> **Exercise 10.** Let A be a set with $|A| = n$. Suppose $x \notin A$ and $y \in A$.
> (a) Prove that $|A \cup \{x\}| = n + 1$.
> (b) Prove that $|A - \{y\}| = n - 1$.

> **Exercise 11.** Let B be a finite set. Prove that if $A \subset B$, then A is finite and $|A| \leq |B|$.

The results of this section now enable us to prove the following theorem. Before looking at the proof, you should read the statement and try to prove it for yourself. While this elementary statement may appear obvious, it is not so easy to prove; in fact, the "counting elements"-type argument you might envision is accomplished by Exercise 11.

Theorem 12. *Suppose $A \subsetneq B$. If B is finite, then A and B do not have the same cardinality.*

Proof. Pick $x \in B - A$. Now $A \subset B - \{x\}$. By Exercise 10(b),

$$|B - \{x\}| = |B| - 1.$$

By Exercise 11,

$$|A| \leq |B - \{x\}| = |B| - 1 < |B|.$$

Hence, $|A| \neq |B|$. By Exercise 9, A and B do not have the same cardinality. □

This theorem and its proof reflect one of the most important ideas in Chapter 3. It is an example of breaking a proof into smaller, easier pieces. These pieces, usually called lemmas, are easier to work on in isolation. Here, the lemmas are the earlier exercises in this section: Exercises 10, 11 and 9.

7.3 Infinite Sets

The definition of infinite sets is perhaps obvious.

Definition 13. A set S is *infinite* if and only if S is not finite.

The next proposition establishes the existence of infinite sets.

Proposition 14. $\mathbb{N}$ *is an infinite set.*

We will give three proofs of this proposition. The first two are similar, while the third is different and will come later (culminating in Supplemental Exercise 6).

Proof #1. Assume that $\mathbb{N}$ is a finite set. Then we can write $\mathbb{N}$ in the form

$$\{n_1, n_2, n_3, \ldots, n_k\}.$$

Hence, $\mathbb{N}$ has a largest element, n_j for some j. Now $n_j + 1 > n_j$ and, therefore, $n_j + 1 \notin \mathbb{N}$. But, since n_j is a natural number, $n_j + 1$ is a natural number. Contradiction! □

You may object to the proof above, asking how we know that the set

$$\{n_1, n_2, n_3, \ldots, n_k\}$$

7.3. Infinite Sets

has a largest element or how we know that $n_j + 1 > n_j$ or, for that matter, how we know that $n_j + 1$ is a natural number. These questions are answered by the construction of the natural numbers in Chapter 9. If you find this proof reasonable now, that is acceptable. Here is a different, but similar proof.

Proof #2. Assume that $\mathbb{N}$ is a finite set. Then we can write $\mathbb{N}$ in the form

$$\{n_1, n_2, n_3, \ldots, n_k\}.$$

We know that $1, 2 \in \mathbb{N}$. The number

$$N = n_1 + n_2 + n_3 + \cdots + n_k$$

is a natural number. However, $N > n_j$ for all j since

$$n_j + 2 > n_j + 1 > n_j$$

and $1, 2$, and n_j are at least two different elements of $\mathbb{N}$. So $N \notin \mathbb{N}$. Contradiction! □

You may have similar objections to Proof #2 as to Proof #1. A different proof is contemplated in Supplemental Exercise 6, based on the following characterization theorem for finite and infinite sets.

Theorem 15. *Let S be a set.*
(a) S is infinite if and only if it has a proper subset of the same cardinality.
(b) S is finite if and only if it has no proper subset of the same cardinality.

As in the proof of Theorem 12, the proof of Theorem 15 is completed in steps. Theorem 12 provided half of part (b); half of part (a) is to be proved in the following exercise.

Exercise 16. Prove that if a set B has a proper subset with the same cardinality, then B is infinite.

Our next goal is to prove the converses of Theorem 12 and Exercise 16, thereby completing the proof of Theorem 15.

Definition 17. A set S is *denumerable* if and only if S and $\mathbb{N}$ have the same cardinality. Sets that are denumerable are also called *countably infinite*.

Definition 18. A set S is *countable* if and only if S is either finite or denumerable.

The question of the existence of a set that is not countable is taken up in Section 7.5.

Since a denumerable set is in one-to-one correspondence with $\mathbb{N}$, we can can think of it as an infinite list indexed by the natural numbers such as

$$\{a_1, a_2, a_3, \ldots\}.$$

Theorem 19. *Every infinite set contains a denumerable subset.*

Proof. Let A be an infinite set. We wish to construct a denumerable subset

$$D = \{a_1, a_2, a_3, \ldots\}$$

by identifying elements $a_1, a_2, \ldots$, in that order. The proof is by induction on the index of the elements. Choose $a_1 \in A$; this is the first step, $n = 1$, of the induction. Next we wish to prove the inductive step: if $D_k = \{a_1, a_2, a_3, \ldots, a_k\} \subset A$, then there exists $a_{k+1} \in A - D_k$. The induction hypothesis is that this is true for $k = n$; we need to show that it is then true for $k = n + 1$. Assume that it is false for $k = n + 1$. That means that $A - D_n = \emptyset$. Hence $A = D_{n+1}$ is finite. This is a contradiction.

That is, we have constructed a sequence of distinct elements of A

$$\{a_1, a_2, a_3, \ldots\},$$

as desired. $\square$

The last sentence of the proof of Theorem 19 looks deceptively simple. The induction actually provides a sequence of finite sets $D_1, D_2, D_3, \ldots$ and not a denumerable set. Mathematicians found an axiom in the eighteenth century that is required in circumstances like this one. It is called the Axiom of Choice and most, but not all, mathematicians use and accept it. We do not wish to study the Axiom of Choice in detail. Suffice it to say that whenever an infinite number of choices are made, the Axiom of Choice is being implemented, unless a universal rule exists to make our infinitely many choices clear. For example, to choose one sock from each of an infinite number of pairs of socks requires the Axiom of Choice; to choose one shoe from each of an infinite number of pairs of shoes does not require it since we can choose all the left shoes.

Having said this, we now present a new proof of Theorem 19, written as most mathematicians might write it.

Second Proof of Theorem 19. Let A be an infinite set. Since A is infinite, A is nonempty. Choose $a_1 \in A$. Now choose $a_2 \in A$ different from a_1; this can be done since $A - \{a_1\}$ is nonempty. Continue in this way: having chosen $a_1, a_2, a_3, \ldots, a_k$ we choose $a_{k+1} \in A - \{a_1, a_2, a_3, \ldots, a_k\}$, which can be done since the difference of a finite set from an infinite set is always nonempty. Continuing this process indefinitely, we construct a denumerable subset $\{a_1, a_2, a_3, \ldots\}$ of A. $\square$

The next exercise requests a proof of a useful lemma. It will be used to prove the converses of Exercise 16 and Theorem 12.

Exercise 20. Suppose D is denumerable and $x \in D$. Prove that $D - \{x\}$ is denumerable.

Exercise 21. Prove that every infinite set has a proper subset of the same cardinality.

Exercise 22. Let S be a set. Prove that if no proper subset of S has the same cardinality as S, then S is finite.

This completes the proof of Theorem 15.

The equivalent definitions of finite and infinite sets, as given in Theorem 15, depend on the Axiom of Choice.

7.4 Countable Sets

We now return to the proof that $\mathbb{N}$ is infinite, Proposition 14. The equivalent definition of infinite in Exercise 21 gives us a different proof that $\mathbb{N}$ is infinite. (See Supplemental Exercise 6.)

7.4 Countable Sets

As we noted, there are two kinds of infinite sets: those that are countable and those that are not countable, that is, uncountable. We will study countable sets next, remembering that countable sets are either finite or denumerable.

Theorem 23. *Every subset of a countable set is countable.*

Proof. Let A be a countable set. We have settled the case where A is finite in Exercise 11. So suppose that A is denumerable. Then we can list the elements of A as, say,

$$A = \{a_1, a_2, a_3, \ldots\}.$$

Let B be a subset of A. Then, using the indices of the list of the elements of A, we let

$$S = \{n \in \mathbb{N} \mid a_n \in B\}.$$

Then $k \mapsto a_k$ defines a one-to-one correspondence from S to B. Either S is finite or infinite. If S is finite, then B is finite and, therefore, countable. Suppose S is infinite. Since we can order the elements of S starting with the smallest, we can label the elements of S

$$n_1, n_2, n_3, \ldots$$

in increasing order. Because $k \mapsto n_k$ defines a one-to-one correspondence from $\mathbb{N}$ to S, we know that S and, therefore, B are denumerable. □

We next ask if all infinite sets are countable? The answer is no. However, we will see that the sets $\mathbb{N}$, $\mathbb{Z}$, and $\mathbb{Q}$ are countable.

Exercise 24. Prove that all denumerable sets have the same cardinality.

Even though $\mathbb{N}$ is a proper subset of $\mathbb{Z}$ and $\mathbb{Z}$ appears to have, roughly, twice as many elements as $\mathbb{N}$, they are shown to have the same cardinality by the following exercise.

Exercise 25. Prove that $\mathbb{Z}$ is denumerable.

An amazing theorem proved by Georg Cantor in 1874 shows that $\mathbb{Q}$, the set of rational numbers, has the same cardinality as $\mathbb{N}$. The proof depends on a diagonal argument.

Theorem 26. *The set of positive rational numbers, $\mathbb{Q}_+^*$, is denumerable.*

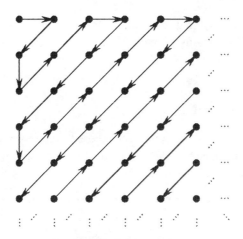

Figure 7.3. Cantor's scheme

Proof. Consider the table of the elements of $\mathbb{Q}_+^*$:

$$\begin{array}{cccccc}
\frac{1}{1} & \frac{2}{1} & \frac{3}{1} & \frac{4}{1} & \frac{5}{1} & \frac{6}{1} \cdots \\
\frac{1}{2} & \frac{2}{2} & \frac{3}{2} & \frac{4}{2} & \frac{5}{2} & \frac{6}{2} \cdots \\
\frac{1}{3} & \frac{2}{3} & \frac{3}{3} & \frac{4}{3} & \frac{5}{3} & \frac{6}{3} \cdots \\
\frac{1}{4} & \frac{2}{4} & \frac{3}{4} & \frac{4}{4} & \frac{5}{4} & \frac{6}{4} \cdots \\
\frac{1}{5} & \frac{2}{5} & \frac{3}{5} & \frac{4}{5} & \frac{5}{5} & \frac{6}{5} \cdots \\
\frac{1}{6} & \frac{2}{6} & \frac{3}{6} & \frac{4}{6} & \frac{5}{6} & \frac{6}{6} \cdots \\
\vdots & \vdots & \vdots & \vdots & \vdots & \ddots
\end{array}$$

We use the pattern in Figure 7.3 on it to create a list of the positive rationals, skipping duplicate rationals in the table as we go along. The list is

$$\mathbb{Q}_+^* = \left\{ 1, 2, \frac{1}{2}, \frac{1}{3}, 3, 4, \frac{3}{2}, \frac{2}{3}, \frac{1}{4}, \frac{1}{5}, 5, 6, \frac{5}{2}, \ldots \right\}.$$

Hence, $\mathbb{Q}_+^*$ is denumerable. □

Exercise 27. Prove that $\mathbb{Q}$ is denumerable.

7.5 Uncountable Sets

We now consider uncountable sets. Yes, they exist; that is, there are different orders of infinity!

Definition 28. A set S is *uncountable* if and only if S is not countable.

7.5. Uncountable Sets

So, every set is exactly one of: finite, denumerable, or uncountable. The proof of the existence of uncountable sets is due to Georg Cantor in 1891.

Theorem 29. *The subset $[0, 1] \subset \mathbb{R}$ is uncountable.*

Proof. Assume that $[0, 1]$ is countable. Then we can list its elements as

$$[0, 1] = \{r_1, r_2, r_3, \ldots\}.$$

Every real number can be written as an infinite decimal; those in $[0, 1]$ can be written with 0 to the left of the decimal point. Therefore, the elements of $[0, 1]$ can be written as

$$r_1 = 0.a_{1,1}a_{1,2}a_{1,3}a_{1,4}a_{1,5}\ldots$$
$$r_2 = 0.a_{2,1}a_{2,2}a_{2,3}a_{2,4}a_{2,5}\ldots$$
$$r_3 = 0.a_{3,1}a_{3,2}a_{3,3}a_{3,4}a_{3,5}\ldots$$
$$r_4 = 0.a_{4,1}a_{4,2}a_{4,3}a_{4,4}a_{4,5}\ldots$$
$$r_5 = 0.a_{5,1}a_{5,2}a_{5,3}a_{5,4}a_{5,5}\ldots$$
$$\ldots$$

where the $a_{i,j}$'s are the jth decimal places of r_i. Let

$$r = 0.b_1b_2b_3b_4b_5\ldots \quad \text{where} \quad b_n = \begin{cases} 3, & \text{if } a_{n,n} \neq 3 \\ 4, & \text{if } a_{n,n} = 3. \end{cases}$$

$r \neq r_n$ for all $n \in \mathbb{N}$ since they differ in the nth decimal place. Therefore, $r \in [0, 1]$, but $r \notin \{r_1, r_2, r_3, \ldots\}$. This is a contradiction. $\square$

In the sense of cardinality, there are more real numbers between any two given rational numbers than there are rational numbers!

Exercise 30. Prove that $\mathbb{R}$ is uncountable.

There are many examples of uncountable sets in Supplemental Exercises 14–20. We will show that all nonempty open intervals (including $\mathbb{R}$) have the same cardinality, and are therefore uncountable.

Exercise 31. Prove that $\mathbb{R}$ and $(0, 1)$ have the same cardinality.

A real number is irrational if it is not rational. That is, the set of irrational numbers is $\mathbb{R} - \mathbb{Q}$.

Exercise 32. Prove that the set of irrational numbers is uncountable.

Exercises 35, 36, and 37 show that uncountable sets arise naturally from countable processes. Next we will define products of infinite collections of sets. In Section 5.2 of Chapter 5, we defined finite products. Actually, we considered only products of two sets and $\mathbb{R}^3$, a product of three sets; this can be generalized to finite products.

A collection, $\mathcal{S}$, of sets is indexed by a set K if

$$\mathcal{S} = \{S_k \mid k \in K\}.$$

In an indexed collection, the elements need not be distinct; that is, an indexed collection may not be a set. For example,

$$\{\mathbb{Z} \mid k \in \mathbb{N}\}$$

is an indexed collection consisting of denumerably many copies of $\mathbb{Z}$.

Definition 33. Suppose $\mathcal{S} = \{S_k \mid k \in K\}$ is a collection of sets indexed by a set K. The *product* of the sets in $\mathcal{S}$ is the set of all functions $f : K \to \cup \mathcal{S}$ with $f(k) \in S_k$ for all $k \in K$.

Suppose $\mathcal{S}$ is a set of sets. So every set in $\mathcal{S}$ appears exactly once. We can then think of the product of the sets in $\mathcal{S}$ as the set of functions $f : \mathcal{S} \to \cup \mathcal{S}$ with $f(S) \in S$ for $S \in \mathcal{S}$. We may think of this as $\mathcal{S}$ being indexed by $\mathcal{S}$ since $\mathcal{S}$ is a set. As long as we distinguish multiple occurrences of a set S in $\mathcal{S}$, you can do this.

For notation, we use

$$\prod_{k \in K} S_k \quad \text{or} \quad \prod_{S \in \mathcal{S}} S$$

to denote the product of the sets in $\mathcal{S}$. Some authors call this a *direct product* and some use the notation

$$\underset{k \in K}{\times} S_k \quad \text{or} \quad \underset{S \in \mathcal{S}}{\times} S.$$

It is one of the most amazing facts in mathematics that the product of infinitely many non-empty sets cannot generally be proven to be non-empty except by using the Axiom of Choice. The notation

$$Y^X = \prod_{x \in X} Y$$

is also commonly used for the set of all functions from X to Y.

Example 34. The collection $\mathcal{S} = \{S_1, S_2\}$ is indexed by $\mathbb{Z}_2$. By definition, $\prod_{k \in \mathbb{Z}_2} S_k$ is the set of all functions $f : \{1, 2\} \to S_1 \cup S_2$ with $f(k) \in S_k$ for both $k \in \mathbb{Z}_2$. That is, for $k = 1, 2$, an element of the product depends on elements $s_1 = f(1) \in S_1$ and $s_2 = f(2) \in S_2$. We can think of this as the ordered pair (s_1, s_2). This is a generic element of $S_1 \times S_2$ according to Definition 1 of Chapter 5.

In light of Example 34, when K is a finite set such as $K = \mathbb{Z}_n = \{1, 2, \ldots, n\}$ we write

$$S_1 \times S_2 \times \cdots \times S_n = \prod_{k \in \mathbb{Z}_n} S_k$$

though this is not technically correct. We ask you to believe that this is not circular: products of collections of sets are defined in terms of functions that are defined in terms of products of two sets.

Exercise 35. Let $S = \{0, 1\}$ and let $P = \prod_{n \in \mathbb{N}} S = S \times S \times S \times \ldots$. ($P$ is a denumerable product of copies of S). Prove that P is uncountable.

7.5. Uncountable Sets

Exercise 36. Prove that the denumerable product of finite sets, each with two or more elements, is uncountable.

Exercise 37. Prove that the power set of $\mathbb{N}$ is uncountable.

Moreover, $\mathcal{P}(\mathbb{N})$ and $\mathbb{R}$ have the same cardinality. The proof of this depends on showing that every element of $(0, 1)$ has a (unique) binary representation, corresponding to a sequence of 0s and 1s (which corresponds to an element of $\mathcal{P}(\mathbb{N})$ as suggested by the hint for Exercise 37).

The following result is often called Cantor's Theorem.

Exercise 38. Let S be a nonempty set. Prove that the power set $\mathcal{P}(S)$ and S do not have the same cardinality.

Remark 39. From what we have done, it is not clear if there exists an uncountable subset of $\mathbb{R}$ that does not have the same cardinality as $\mathbb{R}$. The rejection of this possibility is called the *Continuum Hypothesis*. It turns out that, as is true for the Axiom of Choice, the Continuum Hypothesis is an independent axiom; that is, both its acceptance and its rejection are consistent with our usual set of axioms.

7.6 Supplemental Exercises

Definition Review. There were seven definitions in this chapter. Let S be a set with subset A and let R, $\prec$, $\preceq$, and $\approx$ be relations on S. Define each of the following and give an example of each:

Definition 1. sets A and B have the same cardinality

Definition 8. a set S is *finite*

Definition 13. a set S is *infinite*

Definition 17. a set S is *denumerable*

Definition 18. a set S is *countable*

Definition 28. a set S is *uncountable*

Definition 33. Π is the *product* of the sets in the collection $\mathcal{S} = \{S_k \mid k \in K\}$

Supplemental Exercise 1. Let $A = \{0\}$, $B = \{1, 2, 3\}$, $C = \emptyset$, $D = \{0, -1, -2, -3\}$, $E = \{\emptyset\}$, and $F = \{54, 13, -23\}$. Which pairs of sets have the same cardinality?

Supplemental Exercise 2. Find a number between two distinct real numbers, a and b. Prove that there are infinitely many such.

Supplemental Exercise 3. If for a function $f : \mathbb{N} \to \mathbb{N}$, $f(1) < f(2) < f(3) < \ldots$, prove that $f(n) \geq n$ for all $n \in \mathbb{N}$.

Supplemental Exercise 4. Let $f : A \to B$ be a one-to-one function.
(a) Prove that if B is finite, then A is finite.
(b) Give a counterexample to: if A is finite, then B is finite.

Supplemental Exercise 5. Let $f : A \to B$ be an onto function.
(a) Prove that if A is finite, then B is finite.
(b) Give a counterexample to: if B is finite, then A is finite.

Supplemental Exercise 6. Use Theorem 15 to prove that $\mathbb{N}$ is infinite.

Supplemental Exercise 7. Use Theorem 15 to prove that $\mathbb{Z}$ is infinite.

Supplemental Exercise 8. Use Theorem 15 to prove that $\mathbb{Q}$ is infinite.

Supplemental Exercise 9. Use Theorem 15 to prove that $\mathbb{R}$ is infinite.

Supplemental Exercise 10. Use Theorem 15 to prove that $\mathbb{C}$ is infinite.

Supplemental Exercise 11. Prove that a countable union of countable sets is countable.

Supplemental Exercise 12. Prove that a finite product of countable sets is countable. (Countable products of finite sets need not be countable as is shown in Exercises 35 and 36.)

Supplemental Exercise 13. The *set of algebraic numbers* is the union of the real solution sets of all polynomial equations (in one variable) with integer coefficients. Prove that the set of algebraic numbers is denumerable.

7.6. Supplemental Exercises

Supplemental Exercise 14. Prove that $\mathbb{R}$ and $(0, +\infty)$ have the same cardinality.

Supplemental Exercise 15. Prove that, for a real number a, $(a, +\infty)$ and $(0, +\infty)$ have the same cardinality.

Supplemental Exercise 16. Prove that, for a real number a, $(-\infty, -a)$ and $(a, +\infty)$ have the same cardinality.

Supplemental Exercise 17. Prove that, for a real number a and b with $a < b$, (a, b) and $(0, 1)$ have the same cardinality.

Supplemental Exercise 18. Prove that all nonempty open intervals have the same cardinality.

Supplemental Exercise 19. Prove that $(0, 1)$, $[0, 1)$, $(0, 1]$, and $[0, 1]$ have the same cardinality.

Supplemental Exercise 20. Prove that all nonempty and non-singleton intervals in $\mathbb{R}$ have the same cardinality.

Supplemental Exercise 21. A real number is *transcendental* if it is not algebraic. Prove that the set of transcendental numbers is uncountable.

Supplemental Exercise 22. Prove that a subset of a finite set is finite.

Supplemental Exercise 23. Prove that a subset of $\mathbb{N}$ is countable.

Supplemental Exercise 24. Let A be a finite set with $|A| = n$. Let B consist of subsets of A with cardinality m where $m \in \mathbb{Z}_+$. Prove that B is finite. Find the cardinality of B.

Supplemental Exercise 25. Prove that the set of finite subsets of $\mathbb{N}$ is countable.

Supplemental Exercise 26. Let A and B be finite sets. Prove that the cartesian product $A \times B$ is finite.

Supplemental Exercise 27. Prove that the cartesian product of finitely many countable sets is countable.

Supplemental Exercise 28. Prove that the set of polynomials with integer coefficients is countable.

Supplemental Exercise 29. Prove that a pairwise disjoint collection of open intervals in $\mathbb{R}$ is countable.

Supplemental Exercise 30. Prove that the set of closed intervals $[a, b]$ in $\mathbb{R}$, where a and b are rational numbers, is countable.

Supplemental Exercise 31. Prove that if each of two sets has the same cardinality as a subset of the other one, then the two sets have the same cardinality.

Supplemental Exercise 32. Prove that $\mathbb{N}$ and $\mathbb{N} \times \mathbb{N}$ have the same cardinality.

Supplemental Exercise 33. Prove that for an infinite set A, the set $A \times A$ has the same cardinality as A.

Supplemental Exercise 34. Let A and B be sets. Prove that if A and B have the same cardinality, then $\mathcal{P}(A)$ and $\mathcal{P}(B)$ have the same cardinality.

Supplemental Exercise 35. Let A, B, C, and D be sets. Suppose that A and B have the same cardinality and C and D have the same cardinality. Prove that the sets $A \times C$ and $B \times D$ have the same cardinality.

Supplemental Exercise 36. Prove that $\mathbb{R}$ and $\mathcal{P}(\mathbb{R})$ do not have the same cardinality.

Supplemental Exercise 37. Project. Investigate the history of the Axiom of Choice and explain why it is independent of the usual axioms of set theory.

Supplemental Exercise 38. Project. Investigate the history of the Continuum Hypothesis and explain why it is independent of the usual axioms of set theory.

Supplemental Exercise 39. Project. Investigate cardinal and ordinal arithmetic. Compare and contrast them.

NUMBER SYSTEMS

Algebra of Number Systems

8.1 Introduction: A Road Map

The goal of Part III is to construct some familiar sets of numbers, together with the usual arithmetic operations and the usual ordering. We do this by starting with known concepts from set theory. In Chapter 9, the empty set gives us our starting number, zero. From zero, we use set theoretic tools to construct the natural numbers. This step is abstract and is nearly at the level of axioms.

Zero and the natural numbers lead us to the integers (in Chapter 10) and then to the rational numbers (Chapter 11). The integers and rationals are constructed by equivalence relations on the natural numbers and the integers, respectively.

The construction of the real numbers from the rationals is more difficult. We present two approaches in (Chapters 12 and 13). Along the way, we prove a theorem that characterizes the reals.

Finally, we construct the complex numbers from the reals in Chapter 14. In each construction, except for the complex numbers, a total ordering will be given. We will see that there is no ordering of the complex numbers that extends the properties of the ordering of the real numbers.

8.2 Primary Properties of Number Systems

In each construction, we will also have two operations, addition and multiplication. We have defined what it means for $\leq$ to be a total ordering (or, equivalently, what it means for $<$ to be a strict total ordering) in Section 6.2 of Chapter 6. The following defines what it means to have an operation like addition or multiplication.

Definition 1. A *binary operation* on a set X is a function $\circledast : X \times X \to X$. For $x_1, x_2 \in X$, we denote $\circledast(x_1, x_2)$ by $x_1 \circledast x_2$.

We will call a binary operation an *operation*. Some authors prefer *binary operator* or *operator*. Addition and multiplication are operations.

Remark 2. We always use the symbol $+$ to denote addition. We will follow the usual conventions by writing multiplication in numerous ways:

$$x \cdot y = x \times y = (x)(y) = xy.$$

123

The usual order of operations applies: multiplications have priority over additions. Hence, $x + yz = x + (yz)$ and (usually!) $x + yz \neq (x + y)z$.

Remark 3. The following properties on a set X with addition and multiplication operations:

(A0) Closure of Addition:
   for all $x, y \in X$, $x + y \in X$

(M0) Closure of Multiplication:
   for all $x, y \in X$, $xy \in X$

are called the closure properties. These are often listed as special properties of addition and multiplication. Since we have defined operations as functions from $X \times X$ to X, the closure properties are redundant. That is,

$$+ : X \times X \to X \quad \text{and} \quad \cdot : X \times X \to X$$

are functions. Therefore, if $x, y \in X$, then $(x, y) \in X \times X$ and hence

$$+(x, y) \in X \quad \text{and} \quad \cdot(x, y) \in X.$$

This notation is clumsy and we revert to the usual notation for addition and multiplication:

$$x + y = +(x, y) \quad \text{and} \quad xy = \cdot(x, y).$$

In this chapter and throughout Part III, we will concern ourselves with the algebraic properties of addition and multiplication operations on a totally ordered set X. Let us first compile a list of properties that are to be considered for all $x, y, z \in X$:

(AC) Commutativity of Addition:
   $x + y = y + x$.

(AA) Associativity of Addition:
   $(x + y) + z = x + (y + z)$.

(AZ) Existence of an Additive Identity:
   there exists an element $0 \in X$ such that $0 + x = x + 0 = x$.

(AI) Existence of Additive Inverses:
   there exists an element $-x \in X$ such that $x + (-x) = (-x) + x = 0$.

(MC) Commutativity of Multiplication:
   $xy = yx$.

(MA) Associativity of Multiplication:
   $(xy)z = x(yz)$.

(MU) Existence of a Multiplicative Identity:
   there exists an element $1 \in X$, with $1 \neq 0$, such that $1x = x1 = x$.

(MI) Existence of Multiplicative Inverses:
   if $x \neq 0$, there exists an element $x^{-1} \in X$ such that $xx^{-1} = x^{-1}x = 1$.

(DP) Distributive Properties:
   $(x + y)z = xz + yz$ and $x(y + z) = xy + xz$.

8.2. Primary Properties of Number Systems

This guarantees one 0 and one 1; on the other hand, the inverses depend on the element x.

The letters in parentheses before each property name suggest its name. For example, **(AC)** stands for **A**ddition is **C**ommutative. The additive identity of a field is often called a *zero* and the multiplicative identity of a field is often called a *unity*: **(AZ)** refers to **A**dditive **Z**ero and **(MU)** refers to **M**ultiplicative **U**nity.

Definitions 4 and 8 will present other familiar properties of the real numbers.

Definition 4. A set X with two operations, $+$ and $\cdot$, is a *field* if and only if the properties above are satisfied for all x, y, and z in X.

Remark 5. According to Definition 4, $0 \neq 1$ in a field X. Suppose that $0 = 1$, but all of the other field properties hold. We will show (see Proposition 19) that $0x = 0$ for all $x \in X$. Hence,
$$x = 1x = 0x = 0$$
for all $x \in X$. Therefore, $X = \{0\}$. To avoid this trivial case, most authors include the property that $0 \neq 1$ in their definitions of field, as we have done in **(MU)**.

Remark 6. The Distributive Properties, if written as follows, might be called **Factoring Properties**:
$$xz + yz = (x + y)z \quad \text{and} \quad xy + xz = x(y + z).$$

Remark 7. If X has a commutative multiplication, then the two Distributive Properties are equivalent. In our constructions of $\mathbb{N}$, $\mathbb{Z}$, etc., we will always have a commutative multiplication. Similar comments are in order for identity elements and inverses, as well as for additive properties.

Next, we add an ordering to the mix.

Definition 8. A field X with a total ordering $<$ is an *ordered field* if and only if the following properties are satisfied, for all $x, y, z \in X$:

(PP) Product of Positives:
  if $x > 0$ and $y > 0$, then $xy > 0$

(IX) Additive Cancellation in Inequalities:
  if $x + y < x + z$, then $y < z$

Definition 9. We call x *positive* if and only if $x > 0$. Similarly, we call x *negative* if and only if $x < 0$.

Remark 10. Consider property **(IX)**. It's converse, that $y < z$ implies $x + y < x + z$, may seem obvious. But how do we know this? Perhaps you would say that this is just substitution. We need not appeal to an Axiom of Substitution; in Supplemental Exercise 25, we prove this substitution is implied by two kinds of cancellation properties.

Remark 11. An element z that satisfies
$$z + x = x + z = x$$
for all $x \in X$ is called an *additive identity* of X. Any element y that satisfies $y + x = x + y = 0$ for all $x \in X$ is called an *additive inverse* of x in X. Property **(AZ)** of

a field guarantees the existence of an additive identity and property (**AI**) guarantees the existence of an additive inverse for every element of the field. The uniqueness follows from the existence according to the following two propositions.

Proposition 12. *If a set X with addition has an additive identity, then this identity is unique.*

Proof. Assume that there exist two additive identities 0 and $0'$. Then
$$0' = 0' + 0 = 0.$$
Hence, $0' = 0$. Contradiction! $\square$

The proof of Proposition 12 is a standard proof for many kinds of identity elements. It is sometimes described as "letting the two identities fight it out." The uniqueness of inverses is similarly proved by letting two inverses fight it out. You should compare this with Theorems 42 and 47 of Chapter 5.

Proposition 13. *Suppose X is a set with addition satisfying properties* (**AA**) *and* (**AZ**). *If $x \in X$ has an additive inverse, then it is unique.*

Proof. Assume that $-x$ and x' are distinct additive inverses of x. Then
$$x' = x' + 0 = x' + (x + (-x)) = (x' + x) + (-x) = 0 + (-x) = -x.$$
So $x' = -x$. Contradiction! $\square$

The existence of the multiplicative identity and multiplicative inverses (of nonzero elements) similarly implies their uniqueness.

> **Exercise 14.** Prove that if a set X with multiplication has a multiplicative identity, then it is unique.

> **Exercise 15.** Suppose that X is a set with multiplication satisfying properties (**MA**) and (**MU**). Prove that if $x \in X$ has a multiplicative inverse, then it is unique.

8.3 Secondary Properties

One of the most useful aspects of algebra is that the properties characterize the structures. Moreover, the properties themselves may be used to prove other properties without using direct knowledge about the sets or operations involved.

We will consider a few secondary properties. At first, we will assume that X has only those properties—mostly primary properties—indicated. Later we will simply let X be a field or an ordered field. We could try to give a minimal list of conditions that imply the desired conclusion; however, there may exist two or more distinct minimal lists. That is, a sufficient condition may not be necessary. (See the comments after Exercise 17.)

8.3. Secondary Properties

Remark 16. Suppose that X has an addition operation and that $x, y, z \in X$. It seems reasonable that

$$\text{if } y = z, \text{ then } x + y = x + z.$$

This is a substitution: z is substituted for y. We know that $x = x$ and $y = z$ imply $(x, y) = (x, z)$, which implies, since addition is a function, that $x + y = x + z$. Similarly, if $y = z$, then $y + x = z + x$. These properties are sometimes referred to as **Adding Equals to Equals**.

The converse of Adding Equals to Equals, that $x + y = x + z$ implies $y = z$, is called **Cancellation**; compare it with property **(IX)**. It requires proof.

> **Exercise 17. (AX) Additive Cancellation in Equations:** Suppose X is a set with addition satisfying properties **(AA)**, **(AZ)**, and **(AI)**.
> (a) Prove that $x + y = x + z$ if and only if $y = z$ for all $x, y, z \in X$.
> (b) Prove that $y + x = z + x$ if and only if $y = z$ for all $x, y, z \in X$.

In proving Exercise 17, you probably used additive inverses. Does additive cancellation hold in $\mathbb{N}$? For $y \in \mathbb{N}$, $5 + y = 5 + z$ seems to imply that $y = z$. We will prove this is true in Chapter 9. The proof will be different from your proof of Exercise 17 since there are no additive inverses in $\mathbb{N}$: $-5 \notin \mathbb{N}$. The conditions we give are not, in general, necessary and sufficient.

> **Exercise 18. (MX) Multiplicative Cancellation in Equations:** Suppose X is a set with multiplication satisfying properties **(MA)**, **(MU)**, and **(MI)**.
> (a) Prove that $xy = xz$ if and only if $y = z$ for all $x, y, z \in X$ with $x \neq 0$.
> (b) Prove that $yx = zx$ if and only if $y = z$ for all $x, y, z \in X$ with $x \neq 0$.

In the proof of Exercise 18, one direction of the equivalences is easy; these implications are often referred to as **Multiplying Equals by Equals**.

Property **(MI)** says that nonzero elements have multiplicative inverses. Does 0 have a multiplicative inverse? Can $x^{-1} = 0$? The answers are no according to the next proposition.

Proposition 19. *If X is a set with addition and multiplication satisfying properties* **(AZ)**, **(AX)**, *and* **(DP)**, *then* $0x = 0$ *for all* $x \in X$.

Proof. Since

$$0 + 0x = 0x = (0 + 0)x = 0x + 0x,$$

by additive cancelation, $0 = 0x$. □

Since $1 \neq 0$, 0 has no multiplicative inverse. A multiplicative inverse cannot equal 0. In Remark 5 we saw that if $0 = 1$ were allowed in a field, then 0 would be the only element of the field.

Given an ordering, $<$ or $\leq$, on a set X, we have three transitive relations on X: $<$, $\leq$, and $=$. (By Exercise 12 of Chapter 6, we can always derive $<$ from $\leq$ or vice versa.) The following proposition may seem obvious by substitution, but we can prove it.

Proposition 20. *Suppose X is a set with a strict total ordering $<$ and $x, y, z \in X$. If $x < y$ and $y = z$, then $x < z$.*

Proof. By the definition of $\leq$, $x < y$ implies $x \leq y$ and $y = z$ implies $y \leq z$. Hence, by the transitivity of $\leq$, $x \leq z$.

Assume that $x = z$. By symmetry of $=$, $z = y$. By transitivity of $=$, $x = y$. However, $x < y$, contradicting trichotomy of $<$. Hence, $x \neq z$. Therefore, $x < z$. $\square$

Since the preceding proposition involves an order relation ($<$) and an equivalence relation ($=$), it is not simply a transitive property. Similar properties involving other permutations of $<$, $\leq$, and $=$ are given in Supplemental Exercise 14.

8.4 Isomorphisms and Embeddings

In mathematics, we often study structures that are the same except for the names of the elements, relations, operations, and so on that define them. Such structures are called isomorphic. To understand better what kind of structure we are referring to, here is an example: let $\mathbb{R}$ be the underlying set with the order relation and operations of addition and multiplication providing the structure.

Definition 21. Let X and Y be totally ordered sets together with addition and multiplication. We denote the strict orderings by $<$ and the operations by $+$ and $\times$ for both X and Y. X and Y are *isomorphic* structures if and only if there exists a bijective function $f : X \to Y$, called an *isomorphism*, satisfying

$$x_1 < x_2 \text{ if and only if } f(x_1) < f(x_2),$$
$$f(x_1 + x_2) = f(x_1) + f(x_2),$$
$$f(x_1 \times x_2) = f(x_1) \times f(x_2)$$

for $x_1, x_2 \in X$. An isomorphism from an algebraic structure onto itself is called an *automorphism*.

Remark 22. We call the three conditions in the Definition 21 order-preserving, addition-preserving, and multiplication-preserving, respectively. Collectively, we say that an isomorphism is structure-preserving.

Sometimes we are interested in algebraic structures that do not include an ordering or have only one operation. Then an isomorphism is a bijection that preserves the structure or structures.

The following example and the discussion and exercises following it should illustrate isomorphisms.

Example 23. Let E be the set of even integers. Define a bijective function $f : \mathbb{Z} \to E$ by $f(n) = 2n$. f is order- and addition-preserving since

$$m < n \text{ if and only if } f(m) = 2m < 2n = f(n)$$
$$\text{and } f(m + n) = 2(m + n) = 2m + 2n = f(m) + f(n).$$

f is not multiplication-preserving since $f(1 \cdot 1) = f(1) = 2$ but $f(1) \cdot f(1) = 2 \cdot 2 = 4$. This shows that f is not an isomorphism but it does not show that $\mathbb{Z}$ is not isomorphic to E using some other function as an isomorphism.

The following exercise lists some of the ways that isomorphisms preserve structure.

Exercise 24. Suppose that X and Y are sets with addition and multiplication (no order relations are assumed). Suppose that $f : X \to Y$ is an isomorphism (a bijection that preserves addition and multiplication). Prove:
(a) If 0 is the additive identity in X, then $f(0)$ is the additive identity in Y.
(b) If 1 is the multiplicative identity in X, then $f(1)$ is the multiplicative identity in Y.
(c) If $-x$ is the additive inverse of x in X, then $f(-x)$ is the additive inverse of $f(x)$ in Y.
(d) If x^{-1} is the multiplicative inverse of x in X, then $f(x^{-1})$ is the multiplicative inverse of $f(x)$ in Y.

Such results are useful in showing that structures are not isomorphic.

Example 25. We know that 1 is the multiplicative identity of $\mathbb{Z}$. Does E have a multiplicative identity? Since $1 \notin E$, the answer appears to be no. However, this does not prove it since another element of E could be the multiplicative identity.

Assume $n \in E$ is the multiplicative identity in E. Therefore, $2n = 2$. Since $2 \in E$, $n \in E$, and $E \subset \mathbb{Z}$, we have $2 \in \mathbb{Z}$ and $n \in \mathbb{Z}$. In $\mathbb{Z}$, $2 = 2(1)$. Hence, $2n = 2(1)$ and, by Multiplicative Cancellation, $n = 1$. This implies that $1 \in E$, a contradiction. (We will check that the necessary properties of $\mathbb{Z}$ hold in Chapter 10.)

Exercise 26. Prove that $\mathbb{Z}$ is not isomorphic to E.

Since we think of isomorphic structures as the same, the following idea is natural.

Exercise 27. On a collection of sets with addition and multiplication, "is isomorphic to" is a relation. Prove that this is an equivalence relation.

The equivalence classes for the relation in Exercise 27 are called *isomorphism classes*.

Definition 28. Let X and Y be totally ordered sets together with addition and multiplication. X is *embedded* in Y if and only if there exists $Z \subset Y$ and an isomorphism $f : X \to Z$. The function $g : X \to Y$ defined by $g(x) = f(x)$ is called an *embedding*.

An embedding is a one-to-one function that preserves order, addition, and multiplication.

8.5 Archimedean Ordered Fields

In constructing the number systems, we will check which of the properties of an ordered field are satisfied and whether the least upper bound property is satisfied. As we progress towards the real numbers, the new constructs will possess more properties than their predecessors. This culminates in the construction of the real numbers. The complex numbers form an interesting digression or continuation.

The main goal of Part III, of this text is to prove the following theorem, which is proved in Section 12.5 of Chapter 12.

Theorem 29 (Main Theorem: The Existence and Uniqueness of $\mathbb{R}$). *There exists, up to isomorphism, a unique ordered field with the least upper bound property.*

By "unique up to isomorphism" we mean unique isomorphism class. From the constructions, we will see that $\mathbb{R}$, with the usual ordering and operations, is *the* ordered field with the least upper bound property.

Let us now prove two important properties of ordered fields with the least upper bound property. Let F be an ordered field with the least upper bound property. We denote the sets of *natural, integral,* and *rational field elements of F*, respectively, by

$$F_\mathbb{N} = \{m \in F \mid m = 1 + 1 + \cdots + 1 \text{ for finitely many 1s}\},$$
$$F_\mathbb{Z} = \{m \in F \mid m = 0 \text{ or } m \in F_\mathbb{N} \text{ or } -m \in F_\mathbb{N}\}, \text{ and}$$
$$F_\mathbb{Q} = \{mn^{-1} \mid m \in F_\mathbb{Z} \text{ and } n \in F_\mathbb{N}\}.$$

The following result is usually called the Archimedean Principle.

Theorem 30. *Let F be an ordered field that has the least upper bound property. If $x, y \in F$ and $x > 0$, then there exists $n \in F_\mathbb{N}$ such that $nx > y$.*

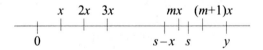

Figure 8.1. The set S in the proof of Theorem 30

Proof. Let $S = \{nx \mid n \in F_\mathbb{N}\}$. S is nonempty since $x \in S$. Assume that $nx \leq y$ for all $n \in F_\mathbb{N}$. Then y is an upper bound of S. Since F has the least upper bound property, S has a least upper bound, s. Now $x > 0$ implies $s - x < s$. Thus, $s - x$ is not an upper bound of S. Hence, there exists $m \in F_\mathbb{N}$ such that $s - x < mx$. Therefore, $s < (m + 1)x$. This contradicts the fact that s is an upper bound of S. □

The Archimedean Principle is closely related to the unboundedness of the natural numbers .

Exercise 31. Let F be an ordered field with the least upper bound property. Prove that $F_\mathbb{N}$ is not bounded.

The following corollary of the Archimedean Principle is a useful and stronger version of it.

Corollary 32. *Let F be an ordered field with the least upper bound property. If $x, y \in F$ and $x > 0$, then there exists $m \in F_\mathbb{Z}$ such that*

$$mx \leq y < (m + 1)x.$$

8.5. Archimedean Ordered Fields

Proof. By the Archimedean Principle, there exists $q \in F_{\mathbb{N}}$ such that $qx > y$. Next, we show that there exists $p \in F_{\mathbb{Z}}$ such that $px < y$. There are two cases: either $y \geq 0$ or $y < 0$.

Case 1. If $y \geq 0$, then $-x < 0 \leq y$. Set $p = -1$ so that $px < y$.
Case 2. If $y \geq 0$, then, by the Archimedean Principle, there exists $n \in F_{\mathbb{N}}$ such that $nx > -y$. Hence, $(-n)x < y$. Set $p = -n$ so that $px < y$.

So, there exist $p, q \in F_{\mathbb{Z}}$ with $p < 0$ and $q > 0$ such that

$$px < y < qx.$$

Let
$$S = \{n \in F_{\mathbb{Z}} \mid n \geq p \text{ and } nx \leq y\}.$$

S is nonempty since $p \in S$. If $n \in S$, then $nx < y < qx$ and hence $n < q$. So q is an upper bound of S and S is a finite set. Since a finite, totally ordered set has a greatest element, let m be the greatest element of S. Since $m \in S$, $mx \leq y$. Since $m + 1 > m$, $m + 1 \notin S$. So $(m+1)x \not\leq y$. Therefore, $mx \leq y < (m+1)x$. □

> **Exercise 33.** Let F be an ordered field with the least upper bound property. If $x, y \in F$ and $x < 0$, then there exists $m \in F_{\mathbb{N}}$ such that $mx \leq y < (m-1)x$.

In words, the following theorem says "the rational field set $F_{\mathbb{Q}}$ is dense in the field F."

Theorem 34. *Let F be an ordered field with the least upper bound property. If $x, y \in F$ and $x < y$, then there exists $r \in F_{\mathbb{Q}}$ such that $x < r < y$.*

Proof. We will use the Archimedean Principle and Corollary 32 to produce $m, n \in F_{\mathbb{N}}$ so that $r = mn^{-1}$.

Since $x < y$, $y - x > 0$. By the Archimedean Principle, there exists $n \in F_{\mathbb{N}}$ such that $n(y - x) > 1$. Thus,

$$nx + 1 < ny. \tag{1}$$

By Corollary 32, there exists $m \in F_{\mathbb{Z}}$ such that $m - 1 \leq nx < m$. Hence,

$$m \leq nx + 1 \quad \text{and} \quad nx < m. \tag{2}$$

The inequalities in (1) and (2) yield

$$nx < m \leq nx + 1 < ny.$$

Therefore $nx < m < ny$ and multiplying by n^{-1} gives:

$$x < mn^{-1} < y.$$

□

8.6 Supplemental Exercises

Definition Review. There were seven italicized terms defined in this chapter. Let S be a set with subset A and let R, $\prec$, $\preceq$, and $\approx$ be relations on S. Define each of the following and give an example of each:

Definition 1. a *binary operation* on a set X

Definition 4. a set X with two operations, $+$ and $\cdot$, is a *field*

Definition 8. a field X with a total ordering $<$ is an *ordered field*

Definition 9. x is *positive*

Definition 9. x is *negative*

Definition 21. totally ordered sets X and Y are *isomorphic*

Definition 28. for totally ordered sets, X is *embedded* in Y

Supplemental Exercise 1. Give reasons for every step in the proof of Proposition 12.

Supplemental Exercise 2. Give reasons for every step in the proof of Proposition 13.

Supplemental Exercise 3. In Proposition 19, we showed that the distributive properties imply $0x = x0 = 0$. Consider the two distributive properties separately. Which implies that $0x = 0$ and which implies $x0 = 0$?

Supplemental Exercise 4. Suppose that $X = \{0\}$. How many different operations can you find on X? Define addition and multiplication on X. Verify that X with this addition and multiplication is a field. What is the multiplicative identity element?

Supplemental Exercise 5. Before continuing beyond this chapter, use your knowledge of the sets $\mathbb{N}$, $\mathbb{Z}_+$, $\mathbb{Z}$, $\mathbb{Q}$, $\mathbb{R}$, and $\mathbb{C}$ to complete the following table. For each set, with the usual addition, multiplication, and ordering, write "Yes" or "No" for each property of an ordered field. For every "No," try to give a reason or counterexample. (For example, $\mathbb{N}$ does not satisfy property **(AZ)**, the existence of an additive identity, since $0 \notin \mathbb{N}$ and $\mathbb{Z}$ does not satisfy property **(MI)**, the existence of multiplicative inverses, since 2 has no multiplicative inverse in $\mathbb{Z}$.)

Property	$\mathbb{N}$	$\mathbb{Z}_+$	$\mathbb{Z}$	$\mathbb{Q}$	$\mathbb{R}$	$\mathbb{C}$
(AC) Commutative Addition						
(AA) Associative Addition						
(AZ) Additive Identity 0						
(AI) Additive Inverses						
(MC) Commutative Multiplication						
(MA) Associative Multiplication						
(MU) Multiplicative Identity 1						
(MI) Multiplicative Inverses						
(DP) Distributive Properties						
(IX) Cancellation in Inequalities						
(PP) Product of Positives						

8.6. Supplemental Exercises

Supplemental Exercise 6. Let X be the set of 2×2 matrices with real entries. Using ordinary addition and multiplication of matrices, determine which of the properties of a field hold. (Check each property for addition, **(AC)**, **(AA)**, **(AZ)**, and **(AI)**, for multiplication, **(MC)**, **(MA)**, **(MU)**, and **(MI)**, and the distributive properties **(DP)**.)

Supplemental Exercise 7. Suppose X has addition and multiplication operations. Prove that if multiplication is commutative, then the First Distributive Property is equivalent to the Second Distributive Property.

Supplemental Exercise 8. Explain why a multiplication operation is closed. Explain why any binary operation is closed.

Supplemental Exercise 9. Prove that in a field X, $(-1)x = x(-1) = -x$ for all $x \in X$.

Supplemental Exercise 10. Suppose that $\mathbb{Z}_3 = \{1, 2, 3\}$ and addition and multiplication are defined by the following tables.

+	1	2	3
1	2	3	1
2	3	1	2
3	1	2	3

·	1	2	3
1	1	2	3
2	2	1	3
3	3	3	3

Prove that $\mathbb{Z}_3$ is a field.

Supplemental Exercise 11. Prove that the only field automorphism of $\mathbb{Z}_3$ from Supplemental Exercise 10 is the identity function, $I_{\mathbb{Z}_3}$.

Supplemental Exercise 12. Suppose that $V = \{0, 1, a, b\}$ and addition and multiplication are defined by the following tables.

+	0	1	a	b
0	0	1	a	b
1	1	0	b	a
a	a	b	0	1
b	b	a	1	0

×	0	1	a	b
0	0	0	0	0
1	0	1	a	b
a	0	a	a	0
b	0	b	0	b

Is V a field? If not, indicate the properties in the definition of a field that fail.

Supplemental Exercise 13. Consider $V = \{0, 1, a, b\}$ with the addition and multiplication operations given in Supplemental Exercise 12. Prove that $f : V \to V$ defined by $f(0) = 0$, $f(1) = 1$, $f(a) = b$, and $f(b) = a$ is an automorphism; i.e., the two operations are preserved.

Supplemental Exercise 14. Suppose X is a set with a partial ordering $\leq$ and $x, y, z \in X$.
(a) Prove that if $x = y$ and $y < z$, then $x < z$.
(b) Prove that if $x < y$ and $y \leq z$, then $x < z$.
(c) Prove that if $x \leq y$ and $y < z$, then $x < z$.
(d) Prove that if $x \leq y$ and $y = z$, then $x \leq z$.
(e) Prove that if $x = y$ and $y \leq z$, then $x \leq z$.

Supplemental Exercise 15. Suppose F is an ordered field. Prove that, for all $x, y \in F$, $x < y$ if and only if $-y < -x$.

Supplemental Exercise 16. Suppose X is an ordered field. Prove that if $0 \neq 1$, then $1 > 0$.

Supplemental Exercise 17. Suppose X is a field. Prove that $-(x + y) = (-x) + (-y)$.

Supplemental Exercise 18. Suppose X is a set with addition satisfying properties (**AA**), (**AZ**), and (**AI**). Prove that, for all $x \in X$,

$$-(-x) = x.$$

Supplemental Exercise 19. Suppose X is a set with multiplication satisfying properties (**MA**), (**MU**), and (**MI**). Prove that, for all $x \in X$ with $x \neq 0$,

$$(x^{-1})^{-1} = x.$$

Supplemental Exercise 20. Suppose that X is a field. Prove that, for all $x, y \in X$, if $xy = 0$, then $x = 0$ or $y = 0$.

Supplemental Exercise 21. In a field X, the following are equivalent for all $x, y \in X$:
If $x \neq 0$ and $y \neq 0$, then $xy \neq 0$.
If $x \neq 0$ and $xy = 0$, then $y = 0$ for all $x, y \in X$.
If $y \neq 0$ and $xy = 0$, then $x = 0$ for all $x, y \in X$.

Supplemental Exercise 22. In a field X, for all $x, y \in X$,

$$xy = 0 \text{ if and only if } x = 0 \text{ or } y = 0.$$

Supplemental Exercise 23. Suppose that X is a field. Prove that, for all $x, y \in X$,

$$(-x)y = x(-y) = -(xy).$$

Supplemental Exercise 24. Suppose that X is a field. Prove that, for all $x, y \in X$,

$$(-x)(-y) = xy.$$

Supplemental Exercise 25. Suppose that X is an ordered field. Prove that, for $x, y, z \in X$,

$$\text{if } y < z, \text{ then } x + y < x + z.$$

Supplemental Exercise 26. Suppose that X is an ordered field. Prove that

$$x > 0 \text{ if and only if } -x < 0 \text{ for all } x \in X.$$

Supplemental Exercise 27. Suppose that X is an ordered field. Prove that

$$\text{if } x > 0 \text{ and } y < z, \text{ then } xy < xz \text{ for all } x, y, z \in X.$$

Supplemental Exercise 28. Suppose that X is an ordered field. Prove that

$$\text{if } x < 0 \text{ and } y < z, \text{ then } xy > xz \text{ for all } x, y, z \in X.$$

Supplemental Exercise 29. Suppose that X is an ordered field. Prove that

$$\text{if } x \neq 0, \text{ then } x^2 = xx > 0 \text{ for all } x \in X.$$

8.6. Supplemental Exercises

Supplemental Exercise 30. Suppose that X is an ordered field. Prove that

$$\text{if } y > x > 0, \text{ then } 0 < y^{-1} < x^{-1} \text{ for all } x, y \in X.$$

Supplemental Exercise 31. Prove that between any two rational numbers there is an irrational number.

Supplemental Exercise 32. Project. This chapter describes the structure and properties of familiar number systems. In abstract (or modern) algebra, the entities on which we define certain "operations" may not be numbers. Investigate how two entities can have "products" and "zero" if they are not numbers. Give examples.

The Natural Numbers

9.1 Introduction

Beginning with this chapter, we make a long journey that terminates in Chapter 14. The work we undertake pertains to the foundations of mathematics, namely to the effort to make explicit and clarify the basic notions, assumptions, and rules upon which rests a large part of the edifice of mathematical knowledge.

We have used numbers (natural, integer, rational, and real) in the previous chapters. A fundamental question is to ask if these objects even exist and what kind of existence can be attributed to them. Our plan is to start with the simplest objects, namely the natural numbers, and then work towards greater and greater complexity, culminating in the set of real numbers and concluding with the complex numbers.

Peano was one of the first mathematicians who contributed towards this goal. Using the axiomatic method, Peano started by postulating the existence of three primitive terms, namely "1," "number," and "successor." Then he gave five axioms:

1. 1 is a number.
2. The successor of any number is a number.
3. No two numbers have the same successor.
4. 1 is not the successor of any number.
5. If 1 has a property and the successor of every number that has the property also has the property, then the property belongs to all numbers.

From the first and second axioms, we obtain an infinite set of numbers. Since 1 is a number and the successor of a number is a number, then the successor of 1 is a number, call it 2. Since 2 is a number and the successor of a number is a number, then the successor of 2 is a number, call it 3. Since 3 is a number and the successor of a number is a number, then the successor of 3 is a number, call it 4, and so on. Thus, we obtain the set

$$\{1, 2, 3, 4, \ldots\}$$

that we call $\mathbb{N}$.

This process has many interpretations and thus it does not define only the set $\mathbb{N}$! For example, take 1 to mean our zero. Then the same process will define the set

$$Z_+ = \{0, 1, 2, 3, 4, 5, \ldots\} = \{0\} \cup \mathbb{N}$$

of nonnegative integers.

Example 1. Let
$$a_n = \frac{1}{2n}$$
for $n = 1, 2, 3, \ldots$. Take "1" to be our $\frac{1}{2}$ and let the "successor" of a term in the sequence represent adding half to the term. Then this sequence is an "interpretation" of Peano's primitive terms and axioms.

> **Exercise 2.** Prove that the process dictated by Peano's primitive terms and axioms has infinitely many interpretations.

The fifth axiom is called the Principle of Induction, which was used in Chapter 3.

9.2 Zero, the Natural Numbers, and Addition

Our approach is not to leave "0" (or "1") as a primitive term, but provide a definition of it. Following von Neumann's set theoretic approach, we define 0 to be the empty set.

Definition 3. Let 0 be $\emptyset$.

Next we give the definition of the immediate successor of a set

Definition 4. The *immediate successor* of a set S is the set $S \cup \{S\}$. It will be denoted by $\mathrm{Succ}(S)$.

In our construction, we will actually start with 0. While we do not regard 0 as a natural number, it will be more natural for our construction to do this. We will construct the set $\{0\} \cup \mathbb{N}$, which we have decided to call $\mathbb{Z}_+$. In the development of numbers, the natural numbers came into use first; the number 0 was not used until much later. The notion of constructing the numbers is far more recent yet; it is due to the work in set theory initiated in the nineteenth century by Georg Cantor and others.

We now define the sets $\mathbb{N}$ and $\mathbb{Z}_+$.

Definition 5. Let 1 be $\mathrm{Succ}(0)$, 2 be $\mathrm{Succ}(1)$, 3 be $\mathrm{Succ}(2)$, 4 be $\mathrm{Succ}(3)$, and so on. The set of numbers constructed in this way yields
$$\mathbb{N} = \{1, 2, 3, 4, 5 \ldots\}$$
and
$$\mathbb{Z}_+ = \{0, 1, 2, 3, 4, 5 \ldots\} = \{0\} \cup \mathbb{N}.$$
We call $\mathbb{N}$ the set of *natural numbers*. We call $\mathbb{Z}_+$ the set of *nonnegative integers*.

Many authors call $\mathbb{Z}_+$ the natural numbers! It does not matter much in practice.
We could compile a list that looks something like the following:
$$0 = \emptyset$$
$$1 = 0 \cup \{0\} = \{\emptyset\}$$
$$2 = 1 \cup \{1\} = \{\emptyset, \{\emptyset\}\}$$
$$3 = 2 \cup \{2\} = \{\emptyset, \{\emptyset\}, \{\emptyset, \{\emptyset\}\}\}$$
$$\ldots$$

9.2. Zero, the Natural Numbers, and Addition

Unfortunately, having done this will not guide us in the development of the natural numbers.

Definition 6. Let $n \in \mathbb{N}$. The *set of successors* of n or *descendancy set* of n is the subset
$$\text{Desc}(n) = \{\,\text{Succ}(n), \text{Succ}(\text{Succ}(n)), \text{Succ}(\text{Succ}(\text{Succ}(n))), \ldots\,\}.$$
For $p \in \mathbb{N}$, we will use the notation $\text{Succ}^p(n)$ to denote the *pth successor* of n, which is
$$\text{Succ}(\ldots(\text{Succ}(n))\ldots),$$
where Succ is taken p times.

$n \notin \text{Desc}(n)$ because if it were there would exist $p \in \mathbb{N}$ such that $n = \text{Succ}^p(n)$.

Definition 7. We define a relation $<$ on $\mathbb{N}$ by $n < m$ if and only if $m \in \text{Desc}(n)$.

> **Exercise 8.** Prove that $<$ is a strict total ordering on $\mathbb{N}$ and, in fact, $\mathbb{N}$ is well-ordered.

Using $<$ and the usual rules for a strict total ordering, we can write $n \leq m$, $m > n$, and $m \geq n$.

Before defining the operation of addition, we consider an inverse operation to the immediate successor.

Definition 9. Let $n \in \mathbb{N}$. The number $m \in \mathbb{Z}_+$ is the *immediate predecessor* of n if and only if the immediate successor of m is n. We denote the immediate predecessor of n by $\text{Pred}(n)$. That is, if the immediate predecessor of n exists, then $n = \text{Succ}(\text{Pred}(n))$.

Lemma 10. *For all $n \in \mathbb{Z}_+$, $n = \text{Pred}(\text{Succ}(n))$.*

Proof. Consider $\text{Pred}(\text{Succ}(n))$. By definition of the immediate predecessor, it is the number whose immediate successor is $\text{Succ}(n)$, that is, n. □

Next we define addition. The definition is inductive.

Definition 11. Let $m, n \in \mathbb{Z}_+$. Define $m + n$ to be
$$m + n = \begin{cases} m, & \text{if } n = 0 \\ \text{Succ}(m + \text{Pred}(n)), & \text{if } n > 0. \end{cases}$$
$m + n$ is called the *sum* of m and n. This defines *addition* on $\mathbb{Z}_+$.

Lemma 12. *For all $n \in \mathbb{Z}_+$ and $p \in \mathbb{N}$, $n + p = \text{Succ}^p(n)$. Moreover,*
$$\text{Desc}(n) = \{\,n+1, n+2, n+3, \ldots\,\}.$$

Proof. We proceed by induction on p. For $p = 1$,
$$n + 1 = \text{Succ}(n + \text{Pred}(1)) = \text{Succ}(n + 0) = \text{Succ}(n).$$
For the inductive step, assume the inductive hypothesis $n + p = \text{Succ}^p(n)$. Then
$$n + \text{Succ}(p) = \text{Succ}(n + \text{Pred}(\text{Succ}(p))) = \text{Succ}(n + p)$$
$$= \text{Succ}(\text{Succ}^p(n)) = \text{Succ}^{p+1}(n).$$
□

> **Exercise 13.** Prove that, for all $m, n \in \mathbb{N}$, $m < n$ if and only if there exists a unique $p \in \mathbb{N}$ such that $m + p = n$.

Let us establish some of the properties of addition in $\mathbb{Z}_+$.

Theorem 14. *Addition on $\mathbb{Z}_+$ is associative.*

Proof. We wish to prove that for all $m, n, p \in \mathbb{Z}_+$,

$$(m + n) + p = m + (n + p).$$

Let m and n be arbitrary elements of $\mathbb{Z}_+$. We proceed by induction on p.

When $p = 0$, by the definition of addition,

$$(m + n) + 0 = m + n \quad \text{and} \quad m + (n + 0) = m + n.$$

Since $=$ is an equivalence relation,

$$(m + n) + 0 = m + (n + 0).$$

For the inductive step, we assume the inductive hypothesis

$$(m + n) + p = m + (n + p).$$

By the definition of addition, Lemma 10, and the induction hypothesis,

$$(m + n) + \text{Succ}(p) = \text{Succ}((m + n) + \text{Pred}(\text{Succ}(p)))$$
$$= \text{Succ}((m + n) + p)$$
$$= \text{Succ}(m + (n + p)).$$

Using the definition of addition twice and Lemma 10 twice,

$$m + (n + \text{Succ}(p)) = \text{Succ}(m + \text{Pred}(n + \text{Succ}(p)))$$
$$= \text{Succ}(m + \text{Pred}(\text{Succ}(n + \text{Pred}(\text{Succ}(p)))))$$
$$= \text{Succ}(m + (n + p)).$$

Using the fact that $=$ is an equivalence relation yields the result. $\square$

We next wish to show that addition on $\mathbb{Z}_+$ is commutative. By the definition of addition, $m + 0 = m$ for $m \in \mathbb{Z}_+$. The next exercise completes the first step in the proof of commutativity.

> **Exercise 15.** Prove that $0 + m = m$ for $m \in \mathbb{Z}_+$.

The following is a corollary.

> **Exercise 16.** Prove that $n = \text{Succ}^n(0)$ for all $n \in \mathbb{N}$.

9.3. Multiplication

With the definition $m + 0 = m$, Exercise 15 shows that 0 is the additive identity element of $\mathbb{Z}_+$. The following exercise justifies use of the definite article when we say "the" additive identity.

Exercise 17. Prove that 0 is the unique additive identity element in $\mathbb{Z}_+$. Moreover, $\mathbb{N}$ does not have an additive identity.

Exercise 18. Prove that addition on $\mathbb{Z}_+$ is commutative.

We will refer to the following as the additive cancellation property for equality.

Exercise 19. Let $m, n, p \in \mathbb{Z}_+$. Prove that $m + n = m + p$ if and only if $n = p$.

Next we consider the relationship between the total ordering $<$ and addition.

Exercise 20. Let $m, n \in \mathbb{Z}_+$. Prove that $m + n = 0$ if and only if $m = n = 0$.

Exercise 21. Let $m, n, p \in \mathbb{Z}_+$. Prove that $m + n < m + p$ implies $n < p$.

We will refer to this as the additive cancellation property for inequality. When we say the cancellation properties for addition, we will mean those for equalities and for inequalities. Of course, the operation may change also!

9.3 Multiplication

Next we develop multiplication in $\mathbb{Z}_+$.

Definition 22. Let $m, n \in \mathbb{Z}_+$. Define mn by

$$mn = \begin{cases} 0, & \text{if } n = 0 \\ m\text{Pred}(n) + m, & \text{if } n > 0. \end{cases}$$

mn is called the *product* of m and n. This defines *multiplication* on $\mathbb{Z}_+$.

We next examine the properties of multiplication. Since the definition of multiplication resembles the distributive property, we start with it.

Exercise 23. Prove that, for $m, n, p \in \mathbb{Z}_+$, $(m + n)p = mp + np$.

Exercise 24. Prove that 1 is the multiplicative identity for $\mathbb{Z}_+$.

Exercise 25. Prove that multiplication on $\mathbb{Z}_+$ is commutative.

From the distributive and commutative properties we get the second distributive property: for all $m, n, p \in \mathbb{Z}_+$, $m(n + p) = mn + mp$.

Exercise 26. Prove that multiplication on $\mathbb{Z}_+$ is associative.

Exercise 27. Let $m, n \in \mathbb{Z}_+$. Prove that $mn = 0$ if and only if $m = 0$ or $n = 0$.

Theorem 28 is the multiplicative cancellation property.

Theorem 28. *Let $m, n, p \in \mathbb{Z}_+$. Prove that if $mn = mp$ and $m \neq 0$, then $n = p$.*

Proof. Assume $n \neq p$. By trichotomy, either $n < p$ or $n > p$.

Suppose that $n < p$. Hence, there exists $q \in \mathbb{N}$ such that $p = n + q$ (by Exercise 13). So
$$mn + 0 = mn = mp = m(n + q) = mn + mq.$$
By additive cancellation, $0 = mq$. By Exercise 27, $m = 0$ or $q = 0$. Since $m \neq 0$, we must have $q = 0$. This contradicts that $q \in \mathbb{N}$.

If $n > p$ we get a similar contradiction. Therefore, $n = p$. □

Exercise 29. Complete the proof of Theorem 28 for $n > p$.

Next we have the other cancellation property for multiplication.

Exercise 30. Let $m, n, p \in \mathbb{Z}_+$ and $m \neq 0$. Prove that $n < p$ if and only if $mn < mp$.

9.4 Supplemental Exercises

Definition Review. There were nine italicized terms defined in this chapter. Let S be a set with subset A and let R, $\prec$, $\preceq$, and $\approx$ be relations on S. Define each of the following and give an example of each:

Definition 3. 0 (i.e., *zero*)

Definition 4. the *immediate successor* of a set S

Definition 5. the set of *natural numbers*

Definition 5. the set of *nonnegative integers*

Definition 6. the *descendancy set* of n

Definition 7. the *order relation* $<$ on $\mathbb{N}$

Definition 9. the *immediate predecessor of n* for $n \in \mathbb{N}$

Definition 11. *addition* in $\mathbb{Z}+$

Definition 22. *multiplication* in $\mathbb{Z}+$

Supplemental Exercise 1. Prove that Pred(0) does not exist.

Supplemental Exercise 2. Prove that the Succ$(n) = n + 1$ for every $n \in \mathbb{Z}_+$ and that Pred$(n) + 1 = n$ for every $n \in \mathbb{N}$.

Supplemental Exercise 3. Prove that addition on $\mathbb{Z}_+$ is closed.

Supplemental Exercise 4. Prove that multiplication on $\mathbb{Z}_+$ is closed.

Supplemental Exercise 5. Prove that, for $m, n, p, q \in \mathbb{Z}_+$,

$$(m + n)(p + q) = mp + mq + np + nq.$$

Supplemental Exercise 6. Prove that $m < mm$ for $m > 2$ without using induction.

Supplemental Exercise 7. Prove that for all $m, n, p, q \in \mathbb{N}$, if $m < n$ and $p < q$, then $m + q < n + q$.

Supplemental Exercise 8. Prove that for all $m, n, p, q \in \mathbb{N}$, if $m < n$ and $p < q$, then $mq < nq$.

Supplemental Exercise 9. Project. Investigate the necessity for the rigorous development of the number system and how it answers the so-called foundational questions of mathematics.

Summary of the Properties of the Nonnegative Integers

The set $\mathbb{Z}_+ = \{0, 1, 2, 3, 4, 5, \ldots\}$, with addition, multiplication, and an order relation, satisfies the following properties, for all $m, n, p \in \mathbb{Z}_+$:

(AC) $m + n = n + m$

(AA) $(m + n) + p = m + (n + p)$

(AZ) 0 is the unique element such that $0 + m = m + 0 = m$

(AX) $m + n = m + p$ if and only if $n = p$ and
$n + m = p + m$ if and only if $n = p$

(MC) $mn = nm$

(MA) $(mn)p = m(np)$

(MU) 1 is the unique element such that $1m = m1 = m$

(MX) for $m \neq 0$, $mn = mp$ if and only if $n = p$ and
for $m \neq 0$, $nm = pm$ if and only if $n = p$

(DP) $(m + n)p = mp + np$ and $m(n + p) = mn + mp$

(PP) if $m > 0$ and $n > 0$ then $mn > 0$

(IX) $n + m < p + m$ if and only if $n < p$

Let $\mathbb{N} = \mathbb{Z}_+^* = \{1, 2, 3, 4, 5, \ldots\}$.

Despite the fact that there are no additive inverses for nonzero elements of $\mathbb{Z}_+$, the **Difference Property** holds: for all $m, n \in \mathbb{Z}_+$,

there exists a unique $d \in \mathbb{N}$ such that $m + d = n$ if and only if $m < n$.

This can be thought of as the definition of the relation $<$. Recall that $m \leq n$ is defined by $m < n$ or $m = n$.

10
The Integers

10.1 Introduction: Integers as Equivalence Classes

In Chapter 9, we constructed the set

$$\mathbb{Z}_+ = \{0, 1, 2, 3, 4, 5, \ldots\}.$$

We call it the set of nonnegative integers. We constructed two operations, addition and multiplication, on $\mathbb{Z}_+$ and a strict order relation $<$. In this chapter, you will not need to refer to the construction in Chapter 9; rather, we have listed on the previous page all of the relevant properties of the construction. These properties of $\mathbb{Z}_+$ are all that we will need in this chapter.

Addition and multiplication are the two operations defined on $\mathbb{Z}_+$. Subtraction and division were never mentioned and in fact neither operation can be defined universally on $\mathbb{Z}_+$. The deficiency of $\mathbb{Z}_+$ with respect to subtraction vis-a-vis addition necessitated its extension to the set of the integers that will be discussed in this chapter. For m and n we want the difference $m - n$ to have a meaning. That is, we want the existence of a number q such that $m = n + q$.

A natural way to proceed would be to attempt to expand $\mathbb{Z}_+$ into $\mathbb{Z}$. One way to do this is to assume that, for $m \in \mathbb{Z}_+$, there exists a number m' such that $m + m' = 0$; we call m' the (additive) inverse of m. Then the set of all integers $\mathbb{Z}$ would be the set of all natural numbers with their inverses and 0.

Setting aside for a moment the problem of defining the new elements, this approach makes it difficult to give even a cumbersome definition of addition. The definition has several cases, which makes the proofs of the basic properties of the operations in $\mathbb{Z}$ tedious.

Therefore, we prefer to start with the following notion. We see that

$$-1 = 0 - 1 = 1 - 2 = 2 - 3 = 3 - 4 = 4 - 5 = \ldots$$

and similar statements can be made for all negative integers and, in fact, for all integers. We can therefore formulate a definition of the integers as equivalence classes of pairs of nonnegative integers.

This construction has the disadvantage that $\mathbb{Z}$ no longer contains $\mathbb{Z}_+$ as we had previously defined it. However, in Section 8.4, we show that $\mathbb{Z}_+$ is embedded in $\mathbb{Z}$; this allows us to think of our previously constructed set $\mathbb{Z}_+$ as a subset of $\mathbb{Z}$.

We start with a definition of a set that we will call $\mathbb{Z}$, the set of integers.

Definition 1. Define a relation $\sim$ on $\mathbb{Z}_+ \times \mathbb{Z}_+$ by $(m,n) \sim (p,q)$ if and only if $m+q = p+n$.

Proposition 2. *The relation $\sim$ on $\mathbb{Z}_+ \times \mathbb{Z}_+$ is an equivalence relation.*

Proof. We show that $\sim$ is reflexive. For $(m,n) \in \mathbb{Z}_+ \times \mathbb{Z}_+$, $(m,n) \sim (m,n)$ if and only if $m+n = m+n$, which is true (by the reflexive property of $=$ on $\mathbb{Z}_+$).

We show that $\sim$ is symmetric. For $(m,n), (p,q)$ in $\mathbb{Z}_+ \times \mathbb{Z}_+$, $(m,n) \sim (p,q)$ if and only if $m+q = p+n$, while $(p,q) \sim (m,n)$ if and only if $p+n = m+q$. The statements $m+q = p+n$ and $p+n = m+q$ are equivalent by the symmetric property of $=$ on $\mathbb{Z}_+$.

It is shown that $\sim$ is transitive in Exercise 3. $\square$

> **Exercise 3.** Complete the proof of Proposition 2.

Definition 4. The set of *integers* is $(\mathbb{Z}_+ \times \mathbb{Z}_+)/\sim$. We denote it as $\mathbb{Z}$. We denote the equivalence class of (m,n) by $[m,n]$, an element of $\mathbb{Z}$.

10.2 A Total Ordering of the Integers

We define a relation on $\mathbb{Z}$ that we show is a total ordering.

Definition 5. Define the relation $<$ on $\mathbb{Z}$ by $[m,n] < [p,q]$ if and only if

$$m+q < p+n.$$

It is important to realize that the symbol $<$ is used in two ways in the definition. The first is a new relation that we define on $\mathbb{Z}$. The second is our old strict total ordering on $\mathbb{Z}_+$. We could use different symbols for each, e.g., use subscripts as in $<_{\mathbb{Z}_+}$ and $<_{\mathbb{Z}}$, but we find these too hideous. There should be no confusion.

As with any operation on equivalence classes, we must check that it is well-defined. That is, we must check that all representatives of the equivalence classes of (m,n) and (p,q) satisfy the defining condition for $<$. Before doing this, we remind you of Proposition 20 in Chapter 8: if $x < y$ and $y = z$, then $x < z$ for all $x, y, z \in X$ and a strict partial ordering $<$ on X.

Proposition 6. *The relation $<$ on $\mathbb{Z}$ is well-defined.*

Proof. Suppose that $[m,n], [p,q] \in \mathbb{Z}$ and $[m,n] < [p,q]$. Thus, $m+q < n+p$. We must show that if $(m',n') \in [m,n]$ and $(p',q') \in [p,q]$, then $m'+q' < n'+p'$. Since $(m',n') \in [m,n]$ and $(p',q') \in [p,q]$,

$$m'+n = m+n' \quad \text{and} \quad p'+q = p+q',$$

respectively. Adding the equations gives us

$$(m'+n) + (p+q') = (m+n') + (p'+q).$$

Using the commutative and associative properties of addition in $\mathbb{Z}_+$ yields

$$(m'+q') + (n+p) = (n'+p') + (m+q). \tag{1}$$

10.3. Addition of Integers

Since $m + q < n + p$,

$$(m' + q') + (m + q) < (m' + q') + (n + p). \tag{2}$$

Now, (1) and (2) imply

$$(m' + q') + (m + q) < (n' + p') + (m + q).$$

By cancellation in $\mathbb{Z}_+$,

$$m' + q' < n' + p'.$$

□

We used only properties (AA), (AC), and (AX) for $\mathbb{Z}_+$ in the proof.

Example 7. We want $-4 < 3$. By our definition of $<$ on $\mathbb{Z}$, we have $[0, 4] < [3, 0]$ in $\mathbb{Z}$ is equivalent to $0 + 0 < 4 + 3 = 7$ in $\mathbb{Z}_+$, as desired. Another example is $-4 < -3$. By our definition of $<$ on $\mathbb{Z}$, we have $[0, 4] < [0, 3]$ in $\mathbb{Z}$ is equivalent to $0 + 3 < 4 + 0$, as desired.

Exercise 8. Prove that $<$ is a strict total ordering on $\mathbb{Z}$.

As for any strict total ordering $<$, there is an associated total ordering $\leq$: define $[m, n] \leq [p, q]$ if and only if $[m, n] < [p, q]$ or $[m, n] = [p, q]$.

Theorem 9. *For $[m, n] \in \mathbb{Z}$, there exists a unique $p \in \mathbb{Z}_+$ such that*

$$[m, n] = \begin{cases} [p, 0], & m \geq n \\ [0, p], & m \leq n. \end{cases}$$

Proof. If $m = n$, then $m + 0 = n + 0 = 0 + n$. Hence, $[m, n] = [0, 0]$ and $p = 0$.

If $m > n$, then by the Difference Property for $\mathbb{Z}_+$ there exists a unique p such that $m = n + p$. Hence, $m + 0 = n + p$ and $[m, n] = [p, 0]$.

If $m < n$, then there exists a unique p such that $n = m + p$. Hence, $m + p = n + 0$ and $[m, n] = [0, p]$. □

We will stop writing $[m, n]$ for an integer and go back to the more usual forms, p or $-p$. We will write p for $[p, 0]$, 0 for $[0, 0]$, and $-p$ for $[0, p]$. Nothing has changed; elements of $\mathbb{Z}$ are still equivalence classes of ordered pairs!

Definition 10. Let $m \in \mathbb{Z}$. m is *positive* if and only if $m > 0$ and m is *negative* if and only if $m < 0$.

10.3 Addition of Integers

We next define the operation of addition on $\mathbb{Z}$. The idea behind the definition is that we want

$$(m - n) + (p - q) = (m + p) - (n + q).$$

Definition 11. Define *addition* in $\mathbb{Z}$ by

$$[m, n] + [p, q] = [m + p, n + q]$$

for $[m, n], [p, q] \in \mathbb{Z}$.

The symbol $+$ is used in two ways in the definition. The first is addition in $\mathbb{Z}$, that is, addition of equivalence classes. The second and third are our old addition in $\mathbb{Z}_+$.

As with any operation on equivalence classes, we must check that it is well-defined. That is, we must check that the equivalence class of the sum $(m + p, n + q)$ is independent of the choice of representatives of the equivalence classes $[m, n]$ and $[p, q]$.

Exercise 12. Prove that addition on $\mathbb{Z}$ is well-defined.

Example 13. We want $2 + (-3) = -1$. 2 is represented in $\mathbb{Z}$ by $(2, 0)$ since $2 - 0 = 2$ and -3 is represented in $\mathbb{Z}$ by $(0, 3)$ since $0 - 3 = -3$. By our definition of addition on $\mathbb{Z}$, we have $[2, 0] + [0, 3] = [(2) + (0), (0) + (3)] = [2, 3] = [0, 1]$. If we had chosen $2 = [5, 3]$ and $-3 = [2, 5]$, we get $[5, 3] + [2, 5] = [7, 8] = [0, 1]$.

Addition in $\mathbb{Z}_+$ is commutative and associative with a unique identity. The corresponding properties for addition in $\mathbb{Z}$ follow.

Proposition 14. (AC): *Addition on $\mathbb{Z}$ is commutative.*

Proof. For $[m, n]$ and $[p, q]$ in $\mathbb{Z}$,

$$[m, n] + [p, q] = [m + p, n + q] \quad \text{and} \quad [p, q] + [m, n] = [p + m, q + n].$$

In $\mathbb{Z}_+ \times \mathbb{Z}_+$, $(m + p, n + q) = (p + m, q + n)$ since $m + p = p + m$ and $n + q = q + n$ by the commutative property of addition on $\mathbb{Z}_+$. □

Exercise 15. (AA): Prove that addition on $\mathbb{Z}$ is associative.

Exercise 16. (AZ): Prove that 0 is the unique additive identity for $\mathbb{Z}$.

For algebraic structures, we usually denote the additive identity element as 0. We know that $0 = [0, 0]$, which contains two meanings of 0 in the previous statement.

We come to the highlight of this chapter: the purpose of defining negative integers was to have additive inverses.

Exercise 17. (AI): Prove that every element of $\mathbb{Z}$ has a unique additive inverse.

The additive inverse of m is denoted by $-m$. We can now define subtraction: $m - n = m + (-n)$ for $m, n \in \mathbb{Z}$.

The proofs of the cancellation properties are much simpler than they were for $\mathbb{Z}_+$.

Exercise 18. (AX): Prove that, for all $m, n, p \in \mathbb{Z}$, $m + n = m + p$ if and only if $n = p$.

10.4. Multiplication of Integers

Commutativity and Exercise 18 imply that, for all $m, n, p \in \mathbb{Z}$, if $n + m = p + m$, then $n = p$.

Proposition 19. (IX): *for $m, n, p \in \mathbb{Z}$, if $m + n < m + p$, then $n < p$.*

Proof. Let $m = [m_1, m_2]$, $n = [n_1, n_2]$, $p = [p_1, p_2]$. The following are equivalent:

$$m + n < m + p$$
$$[m_1, m_2] + [n_1, n_2] < [m_1, m_2] + [p_1, p_2]$$
$$(m_1 + n_1) + (m_2 + p_2) < (m_1 + p_1) + (m_2 + n_2)$$
$$n_1 + p_2 < p_1 + n_2$$

$$[n_1, n_2] < [p_1, p_2]$$
$$n < p$$

It is left to you to explain why these are equivalent. □

Exercise 20. (AX): Prove that, for all $m, n, p \in \mathbb{Z}$, if $m + n = m + p$, then $n = p$.

10.4 Multiplication of Integers

Next we will define multiplication. Since we want

$$(m - n)(p - q) = mp - mq - np + nq,$$

we define multiplication as follows.

Definition 21. Define *multiplication* in $\mathbb{Z}$ by

$$[m, n][p, q] = [mp + nq, np + mq]$$

for $[m, n], [p, q] \in \mathbb{Z}$.

Proposition 22. *Multiplication on $\mathbb{Z}$ is well-defined.*

Proof. Suppose $(m', n') \in [m, n] \in \mathbb{Z}$ and $(p', q') \in [p, q] \in \mathbb{Z}$. Hence, $m' + n = n' + m$ and $p' + q = q' + p$. We will use properties of addition in $\mathbb{Z}_+$. The last two equations give

$$m(q' + p) = m(p' + q)$$
$$n(p' + q) = n(q' + p)$$
$$(n' + m)p' = (m' + n)p'$$
$$(m' + n)q' = (n' + m)q'.$$

By the distributive properties,

$$mq' + mp = mp' + mq$$
$$np' + nq = nq' + np$$
$$n'p' + mp' = m'p' + np'$$
$$m'q' + nq' = n'q' + mq'.$$

Adding these equations give

$$(mq' + mp) + (np' + nq) + (n'p' + mp') + (m'q' + nq')$$
$$= (mp' + mq) + (nq' + np) + (m'p' + np') + (n'q' + mq').$$

By the associative and commutative properties, we have

$$(mp + nq) + (n'p' + m'q') + (mq' + np' + mp' + nq')$$
$$= (np + mq) + (m'p' + n'q') + (mq' + np' + mp' + nq').$$

By the cancellation property,

$$(mp + nq) + (n'p' + m'q') = (np + mq) + (m'p' + n'q'),$$

which implies

$$[m, n][p, q] = [mp + nq, np + mq]$$
$$= [m'p' + n'q', n'p' + m'q']$$
$$= [m', n'][p', q'].$$

Therefore, multiplication is well-defined. □

Example 23. We want $(-2)(5) = -10$. By our definition of multiplication on $\mathbb{Q}$, we have $(-2)(5) = [0, 2][5, 0] = [(0)(5) + (2)(0), (2)(5) + (0)(0)] = [0, 10]$.

Next we prove some properties of multiplication in $\mathbb{Z}$.

Exercise 24. (MC): Prove that multiplication on $\mathbb{Z}$ is commutative.

Exercise 25. (MA): Prove that multiplication on $\mathbb{Z}$ is associative.

Proposition 26. (MU): *Prove that 1 is the unique multiplicative identity for $\mathbb{Z}$.*

Proof. For $m \in \mathbb{Z}$ let $m = [m_1, m_2]$. Then

$$1m = [1, 0][m_1, m_2] = [1m_1 + 0m_2, 0m_1 + 1m_2] = [m_1, m_2] = m.$$

It follows from the commutative property on $\mathbb{Z}$ that $m1 = m$. Uniqueness follows from Exercise 14 of Chapter 8. □

10.5. Embedding the Natural Numbers in the Integers

We again have labeled a multiplicative identity 1. However, $\mathbb{Z}$ lacks multiplicative inverses. To prove this, we will use the following exercises.

Exercise 27. (DP): Prove that the distributive property holds in $\mathbb{Z}$.

Exercise 28. (PP): Prove that if $m > 0$ and $n > 0$, then $mn > 0$.

Alas, $\mathbb{Z}$ still has deficiencies, as the following indicates.

Proposition 29. *2 has no multiplicative inverse in $\mathbb{Z}$.*

Proof. Assume that there is $m \in \mathbb{Z}$ such that $2m = 1$. If $m \leq 0$, then $2m \leq 0$ and, hence, $2m \neq 1$. If $m > 0$, then $2m = (1+1)m = m + m > m \geq 1$. In any case, $2m \neq 1$ and hence 2 has no multiplicative inverse in $\mathbb{Z}$. $\square$

We have one last property on the multiplication of integers.

Exercise 30. (MX): Prove that, for $m, n, p \in \mathbb{Z}$, for $m \neq 0$, $mn = mp$ if and only if $n = p$.

10.5 Embedding the Natural Numbers in the Integers

$\mathbb{N}$ and $\mathbb{Z}_+$, as constructed in Chapter 9, are not subsets of $\mathbb{Z} = (\mathbb{Z}_+ \times \mathbb{Z}_+)/\sim$. This is true since an element of $\mathbb{Z}$ is defined as the equivalence class of a pair of elements of $\mathbb{Z}_+$.

However, this poses no problem, since we can think of $\mathbb{Z}_+$ as a subset of $\mathbb{Z}$ in a natural way. Recall the discussion of isomorphic structures in Section 8.4 of Chapter 8. We have $\mathbb{Z}_+$ and $\mathbb{Z}$ as the underlying sets of our structures and the order relations and operations of addition and multiplication providing their organization of these structures.

Do you see an isomorphic copy of $\mathbb{Z}_+$ in $\mathbb{Z}$? If not, look at Theorem 9. If so, you are now in a position to prove the result in the following exercise.

Exercise 31. Prove that there exists an embedding of $\mathbb{Z}_+$ into $\mathbb{Z}$.

Since isomorphic structures can be thought of as identical, we will say from now on that $\mathbb{Z}_+$ is a subset of $\mathbb{Z}$. To be more precise, we could have called our original $\mathbb{Z}_+$ something else and called its isomorphic image $\mathbb{Z}_+$! That would be functional until we define the rational numbers, when we will have to repeat the process of finding an embedding yet again (and again for real numbers and for complex numbers).

10.6 Supplemental Exercises

Definition Review. There were seven italicized terms defined in this chapter. Define each of the following and give an example of each where practical:

Definition 1. an equivalence relation, $\sim$, on $\mathbb{Z}_+ \times \mathbb{Z}_+$ constructing $\mathbb{Z}$

Definition 4. the set of *integers*

Definition 5. a (strict) total ordering, $<$, on $\mathbb{Z}$

Definition 10. $m \in \mathbb{Z}$ is *positive*

Definition 10. $m \in \mathbb{Z}$ is *negative*

Definition 11. *addition* in $\mathbb{Z}$

Definition 21. *multiplication* in $\mathbb{Z}$

Supplemental Exercise 1. Explain what is wrong with the notation $[-p, 0] \in \mathbb{Z}$. (The statement $[p, 0] + [-p, 0] = 0$ is nonsense!)

Supplemental Exercise 2. Explain what is wrong with this proof of the existence of additive inverses in $\mathbb{Z}$ (Exercise 17): For every $[p, 0] \in \mathbb{Z}$,

$$[p, 0] + [0, p] = [p, p] = [0, 0] = 0.$$

Similarly,

$$[0, p] + [p, 0] = [p, p] = [0, 0] = 0.$$

Supplemental Exercise 3. Prove that $<$ is not a well-ordering on $\mathbb{Z}$.

Supplemental Exercise 4. Let $m \in \mathbb{Z}$. Prove that exactly one of the following holds: m is positive, m is negative, or $m = 0$.

Supplemental Exercise 5. Prove that, for $m \in \mathbb{Z}$, $-(-m) = m$.

Supplemental Exercise 6. Prove that, for $m \in \mathbb{Z}$, $0m = 0$.

Supplemental Exercise 7. Prove that, for $m, n \in \mathbb{Z}$, $m(-n) = (-m)n = -(mn)$.

Supplemental Exercise 8. Prove that if $mn > 0$, then $m > 0$ and $n > 0$, or $m < 0$ and $n < 0$.

Supplemental Exercise 9. Prove that if $m < 0$ and $n < 0$, then $mn > 0$.

Supplemental Exercise 10. Prove that $m > 0$ if and only if $-m < 0$.

Supplemental Exercise 11. Prove that, for $m, n \in \mathbb{Z}_+$, $n < m$ if and only if $n - m \notin \mathbb{N}$.

Supplemental Exercise 12. Project. Investigate the history of negative numbers.

Summary of the Properties of the Integers

The set of integers, $\mathbb{Z}$, with addition ($+$), multiplication ($\cdot$), and a strict total ordering ($<$), satisfies the following properties, for $m, n, p \in \mathbb{Z}$:

(AC) $m + n = n + m$

(AA) $(m + n) + p = m + (n + p)$

(AZ) 0 is the unique element such that $0 + m = m + 0 = m$

(AI) $-m$ is the unique element such that $-m + m = m + (-m) = 0$

(AX) $m + n = m + p$ if and only if $n = p$ and
$n + m = p + m$ if and only if $n = p$

(MC) $mn = nm$

(MA) $(mn)p = m(np)$

(MU) 1 is the unique element such that $1m = m1 = m$

(MX) for $m \neq 0$, $mn = mp$ if and only if $n = p$ and
for $m \neq 0$, $nm = pm$ if and only if $n = p$

(DP) $(m + n)p = mp + np$ and $m(n + p) = mn + mp$

(PP) if $m > 0$ and $n > 0$ then $mn > 0$

(IX) $n + m < p + m$ if and only if $n < p$ and
$n + m < p + m$ if and only if $n < p$

The following properties hold for $m, n, p \in \mathbb{Z}_+$,

Multiplication by 0: $0m = m0 = 0$

Factoring a Minus: $(-m)n = m(-n) = -(mn)$

Negation Reverses Order: $m > 0$ if and only if $-m < 0$

11

The Rational Numbers

11.1 Introduction: Rationals as Equivalence Classes

We now turn to the construction of the rationals. Historically, the positive rationals were used before the negative integers. It seems easier to understand what one half of a melon is than to understand zero melons or minus one melon.

To make division of two integers as universally possible as multiplication, we need an extension of our number system. The construction of $\mathbb{Q}$ from $\mathbb{Z}$ is similar to the construction of $\mathbb{Z}$ from $\mathbb{Z}_+$ in the previous chapter.

For each pair of integers m and n, we require the existence of a unique rational number q (to be viewed as the quotient m/n), such that

$$m = nq.$$

Division by $n = 0$ is not allowed. Suppose $n = 0$. Since any product with 0 in $\mathbb{Z}$ equals 0 we would get

$$m = 0q = 0.$$

By the transitivity of equality on $\mathbb{Z}$, m would have to be 0. This is no good either since

$$0 = 0q$$

has infinitely many solutions in $\mathbb{Z}$. So if anyone asks you "What is 0 divided by 0?" you can answer with any number you like! This may remind you of limits like $\lim_{x \to 0} \frac{kx}{x} = k$ for every $k \in \mathbb{R}$.

Given a rational number q, the integers m and n satisfying the equality $m = nq$, i.e., $\frac{m}{n} = q$, are not uniquely determined. For example, we can write $\frac{1}{2}$ in many ways:

$$\frac{1}{2} = \frac{-1}{-2} = \frac{2}{4} = \frac{-2}{-4} = \frac{3}{6} = \frac{-3}{-6} = \frac{4}{8} = \ldots.$$

Therefore, we define the rationals to be equivalence classes of pairs of integers. So

$$(1, 2), (-1, -2), (2, 4), (-2, -4), (3, 6), (-3, -6), (4, 8), \ldots$$

should be elements of the same equivalence class that will represent $\frac{1}{2}$.

The set of all nonzero integers is denoted $Z^* = \mathbb{Z} - \{0\}$.

155

Definition 1. Define a relation $\sim$ on $\mathbb{Z} \times \mathbb{Z}^*$ by $(m, n) \sim (p, q)$ if and only if $mq = pn$.

Exercise 2. Prove that $\sim$ is an equivalence relation on $\mathbb{Z} \times \mathbb{Z}^*$.

Definition 3. The set of *rational* numbers is the set $(\mathbb{Z} \times \mathbb{Z}^*)/\sim$. We denote this set by $\mathbb{Q}$. The equivalence class of (m, n) is denoted by $[m, n]$; that is, $[m, n]$ is an element of $\mathbb{Q}$.

Example 4. For example, $(1, 2), (-1, -2), (2, 4), \ldots$ are all in the equivalence class $[1, 2]$. We will show that $[1, 2]$ is the multiplicative inverse of 2.

Remark 5 (Important)**.** We could have chosen to define rational numbers differently, allowing only positive n (for the "denominator"). Supplemental Exercises 15 and 16 show that this is possible.

For the remainder of this chapter, when we consider a rational number $[m, n]$, we will assume that $n > 0$. It does not much matter whether you think that $\mathbb{Q} = (\mathbb{Z} \times \mathbb{N})/\sim$ or you think that we choose representatives (m, n) of an equivalence class in $(\mathbb{Z} \times \mathbb{N})/\sim$ with $n \in \mathbb{N}$. That is, we can restrict ourselves *a priori* in the construction of $\mathbb{Q}$ to positive denominators or we can restrict ourselves *a posteriori* to choosing only representatives with positive denominators. See Supplemental Exercise 17.

11.2 A Total Ordering of the Rationals

We define a total ordering on $\mathbb{Q}$. The definition is suggested by

$$\frac{m}{n} < \frac{p}{q} \quad \text{implies} \quad mq < pn.$$

Remark 5 is crucial here: since n and q are positive, this works. If they are not, for example

$$\frac{-1}{2} < \frac{1}{-3} \quad \text{does not imply} \quad (-1)(-3) < (2)(1).$$

Definition 6. Define the relation $<$ on $\mathbb{Q}$ by $[m, n] < [p, q]$ if and only if $mq < pn$.

The symbol $<$ is used in two ways in Definition 6.

Exercise 7. Prove that the relation $<$ on $\mathbb{Q}$ is well-defined.

Example 8. We want $\frac{1}{4} < \frac{1}{3}$. This is equivalent to $\frac{3}{12} < \frac{4}{12}$ and also to $(3)(12) < (4)(12)$. By our definition of $<$ on $\mathbb{Q}$, we have $[1, 4] < [1, 3]$ in $\mathbb{Q}$ is equivalent to $(3)(12) < (4)(12)$, as desired.

Next, we check that $<$ is a strict total ordering on $\mathbb{Q}$.

Exercise 9. Prove that $<$ is a strict total ordering on $\mathbb{Q}$.

Every strict total ordering is trichotomous. We can define a total ordering $\leq$ in terms of $<$ and $=$.

11.3. Addition of Rationals

Let $0 = [0, 1]$; we will see in Exercise 16 that 0 is the unique additive identity. Since we choose $n > 0$, we wish that

$$\frac{m}{n} > \frac{0}{1} \quad \text{implies} \quad m = (m)(1) > (n)(0) = 0.$$

Definition 10. Suppose $[m, n] \in \mathbb{Q}$. $[m, n]$ is *positive* if and only if $m > 0$ and $[m, n]$ is *negative* if and only if $m < 0$.

Instead of writing an element in $\mathbb{Q}$ as the equivalence class of a pair of integers, we can write $q \in \mathbb{Q}$. Nothing has changed: there is a pair (m, n) such that $q = [m, n]$.

11.3 Addition of Rationals

You probably remember that addition of rationals (fractions) is more difficult than multiplication of rationals (fractions). Nevertheless, we start with addition. Following the idea that

$$\frac{m}{n} + \frac{p}{q} = \frac{mq + np}{nq},$$

we next define addition.

Definition 11. Define *addition* in $\mathbb{Q}$ by

$$[m, n] + [p, q] = [mq + np, nq]$$

for $[m, n], [p, q] \in \mathbb{Q}$.

In the definition, the $+$ sign on the left side is understood to denote addition in $\mathbb{Q}$, which differs from addition in $\mathbb{Z}$, which appears on the right side and is denoted by the same symbol.

Since Definition 11 depends on the choice of equivalence class representatives, we must check that addition is well defined.

Exercise 12. Prove that addition on $\mathbb{Q}$ is well defined.

Example 13. We want $\frac{1}{3} + \frac{1}{2} = \frac{5}{6}$. This is equivalent to $\frac{2}{6} + \frac{3}{6} = \frac{5}{6}$. By our definition of addition on $\mathbb{Q}$, we have $[1, 3] + [1, 2] = [(1)(2) + (3)(1), (3)(2)] = [5, 6]$.

Let us establish some properties of addition in $\mathbb{Q}$.

Proposition 14. (AC): *Addition on $\mathbb{Q}$ is commutative.*

Proof. For $[m, n], [p, q] \in \mathbb{Q}$,

$$[m, n] + [p, q] = [mq + np, nq] = [pn + qm, qn] = [p, q] + [m, n].$$

Therefore, addition is commutative. □

Exercise 15. (AA): Prove that addition on $\mathbb{Q}$ is associative.

Exercise 16. (**AZ**): Prove that $0 = [0, 1]$ is the unique additive identity for $\mathbb{Q}$.

Exercise 17. (**AI**): Prove that every element of $\mathbb{Q}$ has a unique additive inverse.

We denote the additive inverse of a number $x \in \mathbb{Q}$ by $-x$. Subtraction is defined as with the integers: for $x, y \in \mathbb{Q}$,

$$x - y = x + (-y).$$

Property (**AX**) is established by the following exercise.

Exercise 18. (**AX**): Prove that the cancellation properties hold for addition in $\mathbb{Q}$:
(a) for all $x, y, z \in \mathbb{Q}$, $x + y = x + z$ if and only if $y = z$ and
(b) for all $x, y, z \in \mathbb{Q}$, $y + x = z + x$ if and only if $y = z$.

11.4 Multiplication of Rationals

Since we want

$$\left(\frac{m}{n}\right)\left(\frac{p}{q}\right) = \frac{mp}{nq},$$

we define multiplication as follows.

Definition 19. Define *multiplication* in $\mathbb{Q}$ by

$$[m, n][p, q] = [mp, nq]$$

for $[m, n], [p, q] \in \mathbb{Q}$.

Since Definition 19 depends on the choice of equivalence class representatives, we must check that multiplication is well-defined.

Exercise 20. Prove that multiplication on $\mathbb{Q}$ is well-defined.

The following exercises establish properties (**MC**), (**MA**), and (**MU**) for multiplication in $\mathbb{Q}$.

Example 21. We want $\left(\frac{1}{2}\right)\left(\frac{6}{5}\right) = \frac{3}{5}$. This is equivalent to

$$\frac{(1)(6)}{(2)(5)} = \frac{6}{10} = \frac{3}{5}.$$

By our definition of multiplication on $\mathbb{Q}$, we have

$$[1, 2][6, 5] = [(1)(6), (2)(5)] = [6, 10] = [3, 5].$$

Exercise 22. (**MC**): Prove that multiplication on $\mathbb{Q}$ is commutative.

Exercise 23. (**MA**): Prove that multiplication on $\mathbb{Q}$ is associative.

11.5. An Ordered Field Containing the Integers

Proposition 24. (**MU**): $[1, 1]$ *is the unique multiplicative identity for* $\mathbb{Q}$.

Proof. For $[m, n] \in \mathbb{Q}$,
$$[1, 1][m, n] = [1m, 1n] = [m, n].$$
Similarly, $[m, n][1, 1] = [m, n]$. The uniqueness follows from Exercise 14 of Chapter 8. □

We set $1 = [1, 1]$. By now you should realize the special nature of the rationals of the form $[m, 1]$.

Set $\mathbb{Q}^* = \mathbb{Q} - \{0\}$. The following exercise establishes the existence of multiplicative inverses in $\mathbb{Q}$.

Exercise 25. (**MI**): Prove that, for each $x \in \mathbb{Q}^*$, there exists a unique multiplicative inverse.

We denote the multiplicative inverse of a number $x \in \mathbb{Q}$ by x^{-1}. Division is defined, for $x \in \mathbb{Q}$ and $y \in \mathbb{Q}^*$, by
$$x \div y = \frac{x}{y} = x/y = xy^{-1}.$$
The following exercise establishes property (**MX**) for multiplication in $\mathbb{Q}$.

Exercise 26. (**MX**): Prove that the following cancellation property holds for multiplication in $\mathbb{Q}$: for $x \in \mathbb{Q}^*$ and $y, z \in \mathbb{Q}$, $xy = xz$ if and only if $y = z$.

11.5 An Ordered Field Containing the Integers

To show that $\mathbb{Q}$ is a field, all that remains is to verify the distributive property (**DP**). First, we give an exercise that will be needed in its proof.

Exercise 27. Prove that $[mp, np] = [m, n]$ in $\mathbb{Q}$ for all $m \in \mathbb{Z}$ and $n, p \in \mathbb{Z}^*$.

While you probably proved Exercise 27 by using the arithmetic of $\mathbb{Z}$, we could argue as follows. We can show that
$$[mp, np] = [m, n][p, p]$$
and that $[p, p] = 1$ (see Supplemental Exercise 12). The result follows from a property of the multiplicative identity. That is, Exercise 27 states that $(x)(1) = x$ for all $x \in \mathbb{Q}$.

Proposition 28. (**DP**): *The following distributive property holds in* $\mathbb{Q}$: *for* $x, y, z \in \mathbb{Q}$, $x(y + z) = xy + xz$.

Proof. Let $x = [a, b]$, $y = [c, d]$, and $z = [e, f]$ in $\mathbb{Q}$. Then
$$\begin{aligned}
x(y + z) &= [a, b]([c, d] + [e, f]) = [a, b][cf + de, df] \\
&= [acf + ade, bdf] = [acf + ade, bdf][b, b] \\
&= [abcf + abde, b^2df] = [(ac)(bf) + (bd)(ae), (bd)(bf)] \\
&= [ac, bd] + [ae, bf] = [a, b][c, d] + [a, b][e, f] \\
&= xy + xz.
\end{aligned}$$

□

Proposition 28 and Supplemental Exercise 21 complete the proof that $\mathbb{Q}$ is a field. We next show that $\mathbb{Q}$ is an ordered field.

Exercise 29. (PP): Suppose $x, y \in \mathbb{Q}$. Prove that $xy > 0$ if and only if either $x > 0$ and $y > 0$, or $x < 0$ and $y < 0$.

Exercise 30. (IX): Prove that cancellation in inequalities holds in $\mathbb{Q}$:
(a) for $x, y, z \in \mathbb{Q}$, $x + y < x + z$ if and only if $y < z$ and
(b) for $x, y, z \in \mathbb{Q}$, $y + x < z + x$ if and only if $y < z$.

This shows that $\mathbb{Q}$ is an ordered field.

We think of $\mathbb{Z}$ as a subset of $\mathbb{Q}$ even though it is not. See Supplemental Exercise 26.

As discussed in Chapter 6, $\mathbb{Q}$ does not have the least upper bound property. We postpone a proof of this until we have constructed the real numbers. Then we will be able to show that $\sqrt{2} \in \mathbb{R}$ and use the denseness of $\mathbb{Q}$ in $\mathbb{R}$ to derive a contradiction from the assumption that the least upper bound property holds for $\mathbb{Q}$.

$\mathbb{Q}$, an ordered field without the least upper bound property, nevertheless satisfies the Archimedean Principle. Compare this with Theorem 30 of Chapter 8.

11.6 Supplemental Exercises

Definition Review. There were seven italicized terms defined in this chapter. Define each of the following and give an example of each where practical:

Definition 1. an equivalence relation, $\sim$, on $\mathbb{Z} \times \mathbb{Z}^*$ constructing $\mathbb{Q}$

Definition 3. the set of *rational* numbers

Definition 6. a (strict) total ordering, $<$, on $\mathbb{Q}$

Definition 10. $q \in \mathbb{Q}$ is *positive*

Definition 10. $q \in \mathbb{Z}$ is *negative*

Definition 11. *addition* in $\mathbb{Q}$

Definition 19. *multiplication* in $\mathbb{Q}$

Supplemental Exercise 1. Explain why $[1, 0]$ is not an element of $\mathbb{Q}$.

Supplemental Exercise 2. Use Definition 3 to verify:

(a) $[1, 2] = [2, 4]$ in $\mathbb{Q}$,

(b) $[-11, -6] = [11, 6]$ in $\mathbb{Q}$,

(c) $[-3, 5] = [3, -5]$ in $\mathbb{Q}$.

Supplemental Exercise 3. Use Definition 3 to:

(a) find $(m, n) \in [5, -7]$ in $\mathbb{Q}$ such that $n > 0$,

(b) find $(p, q) \in [-16, 3]$ in $\mathbb{Q}$ such that $q < 0$,

(c) find $(r, s) \in [-6, -13]$ in $\mathbb{Q}$ such that $s > 0$.

Supplemental Exercise 4. Show that $[0, n] = [0, 1]$ in $\mathbb{Q}$ for all $n \in \mathbb{Z}^*$.

Supplemental Exercise 5. Show that $[2, 3] = [2n, 3n]$ in $\mathbb{Q}$ for all $n \in \mathbb{Z}^*$.

Supplemental Exercise 6. Use Definition 6 to verify:

(a) $[1, 2] < [2, 3]$ in $\mathbb{Q}$,

(b) $[-10, 3] > [-7, 2]$ in $\mathbb{Q}$,

(c) $[-10, 3] > [7, -2]$ in $\mathbb{Q}$. [Be careful!]

Supplemental Exercise 7. Prove that $(m, n) \in \mathbb{Z} \times \mathbb{Z}^*$ represents a positive rational number if and only if $mn > 0$ and that (m, n) represents a negative rational number if and only if $mn < 0$.

Supplemental Exercise 8. Suppose that $m, n \in \mathbb{N}$. Determine which is greater:

(a) $[m, n]$ or $[m, n + 1]$

(b) $[m, n]$ or $[m + 1, n]$

Supplemental Exercise 9. Repeat Supplemental Exercise 8 with $m \in \mathbb{Z}_-^*$ (i.e., $m < 0$) and $n \in \mathbb{N}$.

Supplemental Exercise 10. Suppose that $m, n \in \mathbb{N}$. Under what condition on m and n is it true that $[m, n+1] \leq [m-1, n]$?

Supplemental Exercise 11. Use Definition 11 to compute:

(a) $[2, 3] + [3, 4]$ in $\mathbb{Q}$,

(b) $[-2, 5] + [-7, 3]$ in $\mathbb{Q}$,

(c) $[-4, 8] + [9, 18]$ in $\mathbb{Q}$.

Supplemental Exercise 12. Show that $[n, n] = [1, 1]$ in $\mathbb{Q}$ for all $n \in \mathbb{Z}^*$.

Supplemental Exercise 13. Show that $0 \neq 1$ in $\mathbb{Q}$.

Supplemental Exercise 14. Use Definition 19 to compute:

(a) $[2, 3][3, 4]$ in $\mathbb{Q}$,

(b) $[-2, 5][-7, 3]$ in $\mathbb{Q}$,

(c) $[-4, 8][9, 18]$ in $\mathbb{Q}$.

Supplemental Exercise 15. Prove that for $(m, n) \in \mathbb{Z} \times \mathbb{Z}^*$, $[m, n] = [-m, -n]$.

Supplemental Exercise 16. Prove that for $[m, n] \in \mathbb{Q}$, there exists $(p, q) \in [m, n]$ with $q > 0$.

Supplemental Exercise 17. Prove that $(\mathbb{Z} \times \mathbb{N})/\sim$ is isomorphic to $(\mathbb{Z} \times \mathbb{Z}_*)/\sim$.

Supplemental Exercise 18. Prove that $<$ is not a well-ordering on $\mathbb{Q}$. Prove that $\mathbb{Q}$ has no least element.

Supplemental Exercise 19. Let $q \in \mathbb{Q}$. Prove that exactly one of the following holds: q is positive, q is negative, $q = 0$.

Supplemental Exercise 20. (MX): Prove that the following cancellation property holds for multiplication in $\mathbb{Q}$: for $x \in \mathbb{Q}^*$ and $y, z \in \mathbb{Q}$, $yx = zx$ if and only if $y = z$.

Supplemental Exercise 21. Prove that the following distributive property holds in $\mathbb{Q}$: for $x, y, z \in \mathbb{Q}$, $(y + z)x = yx + zx$.

Supplemental Exercise 22. (PP): Suppose $x, y \in \mathbb{Q}$. Prove that $xy > 0$ if and only if either $x > 0$ and $y > 0$, or $x < 0$ and $y < 0$.

Supplemental Exercise 23. (PP): Suppose $x, y \in \mathbb{Q}$. Prove that if $xy > 0$, then either $x > 0$ and $y > 0$, or $x < 0$ and $y < 0$.

Supplemental Exercise 24. (PP): Suppose $x, y \in \mathbb{Q}$. Prove that $xy > 0$ if $x < 0$ and $y < 0$.

Supplemental Exercise 25. Prove that, for $x, y, z \in \mathbb{Q}$, $y + x < z + x$ if and only if $y < z$.

Supplemental Exercise 26. Prove that there exists an embedding of $\mathbb{Z}$ into $\mathbb{Q}$.

Supplemental Exercise 27. Prove that if $x, y \in \mathbb{Q}$ and $x > 0$, then there exists $n \in \mathbb{N}$ such that $nx > y$. Hence, $\mathbb{Q}$ satisfies the Archimedean Principle

11.6. Supplemental Exercises

Supplemental Exercise 28. Prove that if $x, y \in \mathbb{Q}$ and $x > 0$, then there exists $n \in \mathbb{Z}$ such that $nx \leq y < (n+1)x$. (This is a stronger version of the Archimedean Principle.)

Supplemental Exercise 29. Prove that if $x, z \in \mathbb{Q}$ and $x < z$, then there exists $y \in \mathbb{Q}$ such that $x < y < z$. This is a type of denseness property.

Supplemental Exercise 30. Project. It is said that the invention (discovery?) of irrational numbers created the first crisis in mathematics. Investigate how this crisis was overcome and what was the role of Eudoxus.

Summary of the Properties of the Rationals

The set of rationals, $\mathbb{Q}$, with addition, multiplication, and an order relation is an ordered field. That is, the following properties are satisfied, for all $x, y, z \in \mathbb{Q}$:

(AC) $x + y = y + x$

(AA) $(x + y) + z = x + (y + z)$

(AZ) 0 is the unique element such that $0 + x = x + 0 = x$

(AI) $-x$ is the unique element such that $-x + x = x + (-x) = 0$

(AX) $x + y = x + z$ if and only if $y = z$ and
$y + x = z + x$ if and only if $y = z$

(MA) $xy = yx$

(MA) $(xy)z = x(yz)$

(MU) 1 is the unique element such that $1x = x1 = x$

(MI) x^{-1} is the unique element such that $x^{-1}x = xx^{-1} = 1$

(MX) for $x \neq 0$, $xy = xz$ if and only if $y = z$ and
for $x \neq 0$, $yx = zx$ if and only if $y = z$

(DP) $(x + y)z = xz + yz$ and $x(y + z) = xy + xz$

(PP) if $x > 0$ and $y > 0$, then $xy > 0$

(IX) $y + x < z + x$ if and only if $y < z$ and
$y + x < z + x$ if and only if $y < z$

12

The Real Numbers

12.1 Dedekind Cuts

The set of rationals, $\mathbb{Q}$, has the defect that it does not have the least upper bound property. Another way of looking at this defect is to say that the polynomial $x^2 - 2$ has no (rational) roots. In this chapter we construct the familiar real numbers. The first constructions were given, independently, by Richard Dedekind and Georg Cantor in 1872. In this chapter, we will present the construction due to Dedekind.

We saw in Chapter 6 that $\{x \in \mathbb{Q} \mid 0 < x^2 < 2\}$ does not have a least upper bound in $\mathbb{Q}$. We want a set like this to represent the real number $\sqrt{2}$. What set should represent $-\sqrt{2}$? We modify our original idea somewhat to let

$$\{x \in \mathbb{Q} \mid x < 0 \text{ or } x^2 < 2\} \quad \text{and} \quad \{x \in \mathbb{Q} \mid x < 0 \text{ and } x^2 > 2\}$$

represent $\sqrt{2}$ and $-\sqrt{2}$, respectively. That is, a real number is to be represented by the set of all rational numbers less than it. Then it makes sense to let

$$\{x \in \mathbb{Q} \mid x < q\}$$

represent the rational number q. Let us formalize this.

Definition 1. A *Dedekind cut* is a set $D \subset \mathbb{Q}$ such that:

(1) $D \neq \emptyset$.

(2) $D \neq \mathbb{Q}$.

(3) If $x \in D$, $y \in \mathbb{Q}$ and $y < x$, then $y \in D$.

(4) If $x \in D$, then there exists $y \in D$ such that $x < y$.

A Dedekind cut is also called a *cut* or a *real number*. The set of all cuts is denoted $\mathbb{R}$.

We have defined a cut D to be a nonempty proper subset of $\mathbb{Q}$ with no greatest element that contains all rational numbers less than any number in D. Another reasonable candidate to represent q is

$$D' = \{x \in \mathbb{Q} \mid x \leq q\}.$$

Since we are trying to connect real numbers with the least upper bounds of sets of rational numbers that are bounded above, sets like D' are not useful. The least upper bound of D' is the greatest element, q. Property (4) precludes cuts from having a greatest element.

Throughout this chapter we will keep in mind that real numbers are defined as sets of rational numbers; two real numbers are equal if and only if they are the same subset of $\mathbb{Q}$.

Remark 2 (Important Remark). We will denote rational numbers by lower case letters and cuts by upper case letters.

12.2 Order and Addition of Real Numbers

Before continuing with the construction, let us consider property (3) of a cut D that $x \in D$ and $y < x$ implies $y \in D$. This means that D contains all rational numbers that are less than some rational number in D. Two statements equivalent to this are given in the following exercise.

Exercise 3. Suppose $D \in \mathbb{R}$ and $x, y \in \mathbb{Q}$. Prove that the following are equivalent.
(a) If $x \in D$ and $y < x$, then $y \in D$.
(b) If $x \in D$ and $y \notin D$, then $y > x$.
(c) If $x \notin D$ and $x < y$, then $y \notin D$.

We define a total ordering on $\mathbb{R}$ in the natural way.

Definition 4. Define the relation $<$ on $\mathbb{R}$ by $A < B$ if and only if $A \subsetneq B$.

You may find it more appealing to write the condition as $A \leq B$ if and only if $A \subset B$.

Exercise 5. Prove that $<$ is a total ordering on $\mathbb{R}$.

Exercise 6. Prove that $\mathbb{R}$ has the least upper bound property.

We define addition in a natural way.

Definition 7. Define *addition* on $\mathbb{R}$ by

$$X + Y = \{x + y \mid x \in X \text{ and } y \in Y\}.$$

Unlike our previous constructions, we must prove that $X + Y \in \mathbb{R}$, that is, prove the closure property of addition.

Exercise 8. Prove that addition on $\mathbb{R}$ is closed.

Exercise 9. Prove that addition on $\mathbb{R}$ is commutative.

Exercise 10. Prove that addition on $\mathbb{R}$ is associative.

Exercise 11. Prove that $\{x \in \mathbb{Q} \mid x < 0\}$ is the unique additive identity on $\mathbb{R}$.

12.2. Order and Addition of Real Numbers

As always, we denote the additive identity as 0.

The existence of additive inverses is more difficult to establish. We might be tempted to let the additive inverse of X be the set of all x such that $-x \notin X$. This does not work since this set may have a greatest element. For example,

$$X = \{x \in \mathbb{Q} \mid x < 0\}.$$

The set

$$\{x \in \mathbb{Q} \mid -x \notin X\} = \{x \in \mathbb{Q} \mid -x \not< 0\}$$
$$= \{x \in \mathbb{Q} \mid x \leq 0\}$$

has 0 as its greatest element and hence is not a cut.

We define $-X$ as

$$\{x \in \mathbb{Q} \mid -x \notin X \text{ and } -x \text{ is not the least element of } \mathbb{Q} - X\}$$
$$= \{x \in \mathbb{Q} \mid \text{there exists } y > 0 \text{ such that } -x - y \notin X\}.$$

We will show that $-X \in \mathbb{R}$, and $X + (-X) = 0$ or $-X + X = 0$.

Lemma 12. *For $X \in \mathbb{R}$, $-X \in \mathbb{R}$.*

Proof. Recall the four properties in Definition 1 of a cut. By property (2), $X \neq \mathbb{Q}$. So there exists $y \in \mathbb{Q}$ such that $y \notin X$. Set $x = -y - 1$. Hence, $y = -x - 1 = -(x+1) \notin X$. Since $-(x+1) < -x$, $-x \notin X$ and $x \in -X$. This proves property (1), $-X \neq \emptyset$.

If $y \in X$, then $-y \notin -X$. This proves property (2), $-X \neq \mathbb{Q}$.

Choose $x \in -X$ and $w > 0$ such that $-x - w \notin X$. Suppose $y < x$. Then $-y > -x$ and $-y - w > -x - w$. Hence $-y - w \notin X$. Therefore, $y \in -X$ and property (3) is proved.

For x and w as in the previous paragraph, $z = x + \frac{w}{2} > x$ and $-z - \frac{w}{2} = -x - w \notin X$. Therefore, $z \in -X$ and property (4) is proved. □

Lemma 13. *For $X \in \mathbb{R}$, $X + (-X) = 0$.*

Proof. Let $x \in X$ and $y \in -X$. Thus, $-y \notin X$. By Exercise 3, $-y > x$. Therefore, $x + y < 0$ and $X + (-X) \leq 0$.

Choose $v \in 0$. Let $w = -\frac{v}{2}$. Since $v < 0$, $w > 0$. By Supplemental Exercise 28 of Chapter 11, a strong version of the Archimedean Property of $\mathbb{Q}$, there exists $n \in \mathbb{Z}$ such that $nw \in X$ but $(n+1)w \notin X$. Set $x = nw$ and $y = -(n+2)w$. Since $-y - w = (n+2)w - w = (n+1)w \notin X$, $y \in -X$. Now $x + y = nw - (n+2)w = -2w = v$. Therefore, $0 \leq X + (-X)$.

Combining the results gives us $X + (-X) = 0$. □

> **Exercise 14.** Complete the proof that every element of $\mathbb{R}$ has a unique additive inverse.

Having proved the properties for addition, we know from Chapter 8 that several secondary properties hold.

> **Exercise 15.** Prove that cancellation in inequalities holds: $X + Y < X + Z$ if and only if $Y < Z$ for $X, Y, Z \in \mathbb{R}$.

12.3 Multiplication of Real Numbers

Defining multiplication in $\mathbb{R}$ is a little less natural. First we define the product of positive real numbers. We use rules of signs to define products involving negative real numbers.

Definition 16. Let $X, Y \in \mathbb{R}$. If $X > 0$ and $Y > 0$, then

$$XY = \{z \in \mathbb{Q} \mid z \leq 0 \text{ or there exist } x \in X \text{ and } y \in Y$$
$$\text{such that } x > 0, y > 0, \text{ and } z = xy\}.$$

Otherwise, define the product as follows:

$$XY = \begin{cases} 0, & \text{if } X = 0 \text{ or } Y = 0 \\ -[(-X)Y], & \text{if } X < 0 \text{ and } Y > 0 \\ -[X(-Y)], & \text{if } X > 0 \text{ and } Y < 0 \\ (-X)(-Y), & \text{if } X < 0 \text{ and } Y < 0. \end{cases}$$

We will prove the primary properties for multiplication on the set

$$\mathbb{R}_+^* = \{X \in \mathbb{R} \mid X > 0\}$$

of positive real numbers first.

> **Exercise 17.** Prove that multiplication on $\mathbb{R}_+^*$ is closed.

This proves the product of positives property: if $X > 0$ and $Y > 0$, then $XY > 0$ for all $X, Y \in \mathbb{R}$.

> **Exercise 18.** Prove that multiplication on $\mathbb{R}_+^*$ is commutative.

> **Exercise 19.** Prove that multiplication on $\mathbb{R}_+^*$ is associative.

> **Exercise 20.** Prove that $\{x \in \mathbb{Q} \mid x < 1\}$ is the multiplicative identity on $\mathbb{R}_+^*$.

As always, we denote the multiplicative identity as 1.

> **Exercise 21.** Prove that every $X \in \mathbb{R}_+^*$ has a multiplicative inverse.

Next we prove the primary properties for multiplication on $\mathbb{R}$.

> **Exercise 22.** Prove that multiplication on $\mathbb{R}$ is closed.

> **Exercise 23.** Prove that multiplication on $\mathbb{R}$ is commutative.

> **Exercise 24.** Prove that multiplication on $\mathbb{R}$ is associative.

12.4. Embedding the Rationals in the Reals

Exercise 25. Prove that 1 is the multiplicative identity on $\mathbb{R}$.

Exercise 26. Prove that $X \in \mathbb{R}$, $X \neq 0$ has a multiplicative inverse.

As always, we denote the multiplicative inverse of X as X^{-1}. We have now shown that $\mathbb{R}$ is an ordered field.

12.4 Embedding the Rationals in the Reals

We have not defined $\mathbb{R}$ to extend $\mathbb{Q}$ but $\mathbb{Q}$ is naturally embedded in $\mathbb{R}$. By relabeling, we can think of $\mathbb{Q}$ as a subset of $\mathbb{R}$.

Exercise 27. Prove that there exists an embedding of $\mathbb{Q}$ into $\mathbb{R}$.

Collecting the results of this chapter, we have proved the following theorem. (A *subfield* of a field is a subset that is a field under the operations on the ambient set.)

Theorem 28. $\mathbb{R}$ *is an ordered field with the least upper bound property and with $\mathbb{Q}$ as a subfield.*

From Theorem 30 of Chapter 8, we know that $\mathbb{R}$ satisfies the Archimedean Principle and from Theorem 34 of Chapter 8, we know that $\mathbb{Q}$ is dense in $\mathbb{R}$. That is, for all $x, y \in \mathbb{R}$,

if $x > 0$, then there exists $n \in \mathbb{N}$ such that $nx > y$ and

if $x < y$, then there exists $q \in \mathbb{Q}$ such that $x < q < y$.

12.5 Uniqueness of the Set of Real Numbers

We will show that the properties of $\mathbb{R}$ that we have found completely characterize $\mathbb{R}$: an X with these properties is isomorphic to $\mathbb{R}$. This is the Main Theorem mentioned in Section 8.5 of Chapter 8.

Theorem 29 (Main Theorem). *There exists, up to isomorphism, a unique ordered field with the least upper bound property.*

The proof of the Main Theorem is long, so we will break it into eight steps. Suppose F is an ordered field with the least upper bound property. We will write everything having to do with F—its elements, order relation, and operations—in **bold**. We will prove Theorem 29 by constructing an isomorphism $f : \mathbb{R} \to \boldsymbol{F}$. Here is an outline of the construction:

Step 1: construct a function $f_\mathbb{Z} : \mathbb{Z} \to \boldsymbol{F}$

Step 2: extend $f_\mathbb{Z}$ to a function $f_\mathbb{Q} : \mathbb{Q} \to \boldsymbol{F}$

Step 3: extend $f_\mathbb{Q}$ to a function $f : \mathbb{R} \to \boldsymbol{F}$

Step 4: show that f preserves order

Step 5: show that f is one-to-one

Step 6: show that f is onto

Step 7: show that f preserves addition

Step 8: show that f preserves multiplication

Proof of the Main Theorem.
STEP 1: Define $f_{\mathbb{Z}}$ inductively by $f_{\mathbb{Z}}(0) = \mathbf{0}$ and

$$f_{\mathbb{Z}}(n) = \begin{cases} \overbrace{\mathbf{1} + \cdots + \mathbf{1}}^{n \text{ times}}, & \text{if } n > 0 \\ -(\underbrace{\mathbf{1} + \cdots + \mathbf{1}}_{|n| \text{ times}}), & \text{if } n < 0. \end{cases}$$

Let us give some notation for the n-fold sum and its negation:

$$\mathbf{n} = \mathbf{1} + \cdots + \mathbf{1} \quad \text{and} \quad -\mathbf{n} = -(\mathbf{1} + \cdots + \mathbf{1}).$$

Exercise 30. Prove that $f_{\mathbb{Z}} : \mathbb{Z} \to F$ preserves operations.

STEP 2: We will write rational numbers in the form m/n where $m \in \mathbb{Z}$ and $n \in \mathbb{N}$. (E.g., $n = n/1$.) We extend the function $f_{\mathbb{Z}}$ of Step 1 to $f_{\mathbb{Q}} : \mathbb{Q} \to F$ by defining

$$f_{\mathbb{Q}}(m/n) = \mathbf{m}\,\mathbf{n}^{-1}.$$

By the following exercise, the right-hand side is defined.

Exercise 31. Prove that if $n \in \mathbb{N}$, then $f_{\mathbb{Q}}(n) = f_{\mathbb{Z}}(n) = \mathbf{n} \neq \mathbf{0}$.

The function $f_{\mathbb{Q}}$ is well-defined since

$$\frac{m}{n} = \frac{p}{q} \implies mq = np \implies \mathbf{m}\mathbf{q} = \mathbf{n}\mathbf{p} \implies \mathbf{m}\mathbf{n}^{-1} = \mathbf{p}\mathbf{q}^{-1},$$

by field properties of $\mathbb{Q}$ and F and Exercise 30. $f_{\mathbb{Q}}$ extends $f_{\mathbb{Z}}$ since $m \in \mathbb{Z}$ implies

$$f_{\mathbb{Q}}(m/1) = \mathbf{m}\mathbf{1}^{-1} = \mathbf{m}\mathbf{1} = \mathbf{m} = f_{\mathbb{Z}}(m).$$

Exercise 32. Prove that $f_{\mathbb{Q}} : \mathbb{Q} \to F$ preserves operations.

STEP 3: Let $x \in \mathbb{R}$. Thinking of x as a cut,

$$x = \{q \in \mathbb{Q} \mid q < x\}.$$

Extend the function $f_{\mathbb{Q}}$ of Step 2 to $f : \mathbb{R} \to F$ by

$$f(x) = \sup f_{\mathbb{Q}}(\{q \in \mathbb{Q} \mid q < x\})$$
$$= \sup \{f_{\mathbb{Q}}(q) \mid q \in \mathbb{Q} \text{ and } q < X\}.$$

If the least upper bound exists, it is unique and f is well-defined.

Exercise 33. Prove that $\sup f_{\mathbb{Q}}(\{q \in \mathbb{Q} \mid q < x\})$ exists.

12.5. Uniqueness of the Set of Real Numbers

From Section 8.5 of Chapter 8, the definitions of the sets $F_{\mathbb{N}}$, $F_{\mathbb{Z}}$, and $F_{\mathbb{Q}}$ of natural, integral, and rational field elements of F, respectively.

Exercise 34. Prove that $f_{\mathbb{Q}}(\mathbb{N}) = F_{\mathbb{N}}$, $f_{\mathbb{Q}}(\mathbb{Z}) = F_{\mathbb{Z}}$, and $f_{\mathbb{Q}}(\mathbb{Q}) = F_{\mathbb{Q}}$.

We next prove that f extends $f_{\mathbb{Q}}$, that is, for $q \in \mathbb{Q}$, $f(q) = f_{\mathbb{Q}}(q)$. Since, by Exercise 32, $f_{\mathbb{Q}}(p) < f_{\mathbb{Q}}(q)$ for $p < q$, we know that

$$f(q) = \sup\{f_{\mathbb{Q}}(p) \mid p \in \mathbb{Q} \text{ and } p < q\} \leq f_{\mathbb{Q}}(q).$$

Assume that $f(q) < f_{\mathbb{Q}}(q)$. Since $F_{\mathbb{Q}}$ is dense in F (Theorem 34 of Chapter 8), by Exercise 34, there exists $r \in \mathbb{Q}$ such that

$$f(q) < f(r) < f_{\mathbb{Q}}(q).$$

Exercise 35. Complete the proof that f extends $f_{\mathbb{Q}}$ by deriving a contradiction.

STEP 4:

Exercise 36. Prove that f preserves order.

STEP 5:

Exercise 37. Prove that f is one-to-one.

STEP 6:

Exercise 38. Prove that f is onto.

STEP 7:

Exercise 39. Prove that f preserves addition.

STEP 8:

Exercise 40. Prove that f preserves multiplication.

Since we have shown that every ordered field with the least upper bound property is isomorphic to $\mathbb{R}$, Theorem 29 follows because isomorphism is an equivalence relation. □

The construction of the isomorphism above may seem a bit contrived to you at first. Nothing could be further from the truth. It is natural and unique, as the following exercises demonstrate.

Exercise 41. Prove that the only isomorphism of $\mathbb{R}$ onto itself is the identity function.

The result of Exercise 41 can be stated by saying that the identity function $I_{\mathbb{R}}$ is the unique (ordered) field automorphism of $\mathbb{R}$.

Exercise 42. Let F be an ordered field with the least upper bound property. Prove that the f is the unique isomorphism from $\mathbb{R}$ onto F.

12.6 Supplemental Exercises

Definition Review. There were four definitions in this chapter. Let S be a set with subset A and let R, $\prec$, $\preceq$, and $\approx$ be relations on S. Define each of the following and give an example of each:

Definition 1. a *Dedekind cut*

Definition 4. the order relation $<$ on $\mathbb{R}$

Definition 7. *addition* in $\mathbb{R}$

Definition 16. *multiplication* in $\mathbb{R}$

Supplemental Exercise 1. Project. Suppose we say that a real number is a number of the form

$$a.a_1 a_2 a_3 a_4 \ldots.$$

This may involve infinite sums. How would you define addition and multiplication of real numbers so defined?

13

Cantor's Reals

13.1 Convergence of Sequences of Rational Numbers

In Chapter 12, we constructed the set of real numbers from the set of rational numbers using Dedekind cuts. We saw that the cuts completed the rationals with respect to the least upper bound property. In this chapter, we will give a construction devised by Cantor of the reals from the rationals using Cauchy sequences. The reals complete the rationals. The construction by cuts is algebraic and the construction by Cauchy sequences is analytic.

A sequence is a denumerable list of numbers, not necessarily distinct. We restrict our attention to sequences of rational or real numbers, that is, sequences in $\mathbb{Q}$ or $\mathbb{R}$.

Definition 1. Let X be a set. A *sequence* in X is a function from $\mathbb{N}$ to X.

Rarely is a sequence denoted as a function $f : \mathbb{N} \to \mathbb{Q}$. Usually it will be denoted as either

$$\{q_n\} \quad \text{or} \quad \{q_1, q_2, q_3, q_4, \ldots\}.$$

The connection between f and the notation is that $q_n = f(n)$ for $n \in \mathbb{N}$. The notation for sequences looks like a set. This is misleading since terms in the sequence may repeat.

Example 2. $A = \{n\}$ and $B = \{1, 1, 1, 1, \ldots\}$ denote two sequences in $\mathbb{Q}$. The range of A is $\mathbb{N}$ and the range of B is $\{1\}$. We could also write the sequences as $A = \{1, 2, 3, 4, \ldots\}$ and $B = \{1\}$.

Our primary concern will be the convergence of sequences, which does not happen in general.

Definition 3. A sequence $\{q_n\}$ in $\mathbb{Q}$ (respectively, in $\mathbb{R}$) *converges* to q in $\mathbb{Q}$ (respectively, in $\mathbb{R}$) if and only if for every $\varepsilon > 0$ there exists a natural number N such that

$$|q_n - q| < \varepsilon$$

whenever

$$n \geq N.$$

A sequence $\{q_n\}$ in $\mathbb{Q}$ (respectively, in $\mathbb{R}$) is *convergent* if there is $q \in \mathbb{Q}$ (respectively, in $\mathbb{R}$) such that $\{q_n\}$ converges to q.

The idea is that a sequence $\{q_n\}$ in $\mathbb{Q}$ converges to the *limit* q if the terms q_n get closer and closer to q as n gets larger, that is, as you go further into the endless *tail* of the sequence. ε measures how close the terms gets to q and N measures how far into the sequence you have to go to guarantee that the terms stay within ε of q.

A sequence in $\mathbb{Q}$ is also a sequence in $\mathbb{R}$. We will see that a sequence in $\mathbb{Q}$ may converge in $\mathbb{R}$ and fail to converge in $\mathbb{Q}$.

Example 4. (a) The sequence $\{0\}$ converges to 0 since the difference of its terms from 0 is always 0.

(b) The sequence $\{1/n\}$ converges to 0 since $\frac{1}{n}$ can be made as small as we wish by choosing n large enough. This is the Archimedean Property of $\mathbb{Q}$: for a positive $\varepsilon \in \mathbb{Q}$ there exists $n \in \mathbb{N}$ such that $n\varepsilon > 1$ (that is, $0 < \frac{1}{n} < \varepsilon$).

(c) The sequence

$$\{2, 1.5, 1.42, 1.415, 1.4143, 1.41422, 1.414214, \ldots\},$$

whose terms are a rational approximation of $\sqrt{2}$, does not converge in $\mathbb{Q}$.

The goal of the construction in this chapter is to make sequences like the one in the last example convergent in the larger set; that is, convergent to $\sqrt{2} \in \mathbb{R}$.

Before continuing with sequences, we must prove an important inequality, the Triangle Inequality. It will be useful for us since it deals with expressions like $|q_n - q|$ in the definition of convergent sequence.

Exercise 5. Prove that, for all $p, q \in \mathbb{R}$,

$$|p + q| \leq |p| + |q|.$$

Moreover, $|p + q| \geq |p| - |q|$.

Properties like the following about convergent sequences can be useful.

Exercise 6. Suppose $\{x_n\}$ is a sequence in $\mathbb{R}$ that converges to x and $\{y_n\}$ is a sequence in $\mathbb{R}$ that converges to y. Prove that $\{x_n + y_n\}$ converges to $x + y$.

Exercise 7. Suppose $\{x_n\}$ is a sequence in $\mathbb{R}$ that converges to x and $\{y_n\}$ is a sequence in $\mathbb{R}$ that converges to y. Prove that $\{x_n - y_n\}$ converges to $x - y$.

Exercise 8. Suppose $\{x_n\}$ is a sequence in $\mathbb{R}$ that converges to x and $\{y_n\}$ is a sequence in $\mathbb{R}$ that converges to y. Prove that $\{x_n y_n\}$ converges to xy.

Exercise 9. Suppose $\{x_n\}$ is a sequence in $\mathbb{R}$ that converges to x and $\{y_n\}$ is a sequence in $\mathbb{R}$ that converges to y. Prove that $\{x_n/y_n\}$ converges to x/y provided that $y_n \neq 0$ for $n \in \mathbb{N}$ and $y \neq 0$.

13.2 Cauchy Sequences of Rational Numbers

What is it about convergent sequences that makes them convergent? A convergent sequence requires the existence of a limit q in $\mathbb{Q}$ or $\mathbb{R}$ for it to converge to. It also must behave so that q is the number to which it converges. If the terms q_n get closer and closer to q, then they also get closer and closer to each other. The following exercise should help to explain this.

> **Exercise 10.** Suppose that $\{q_n\}$ is a convergent sequence in either $\mathbb{Q}$ or $\mathbb{R}$. Prove that for every $\varepsilon > 0$ there exists a natural number N such that $|q_m - q_n| < \varepsilon$ whenever $m \geq N$ and $n \geq N$.

Sequences with the property in Exercise 10 above were first studied by Cauchy.

Definition 11. A sequence $\{q_n\}$ in $\mathbb{Q}$ or in $\mathbb{R}$ is called a *Cauchy* sequence if and only if for every $\varepsilon > 0$ there exists a natural number N such that $|q_m - q_n| < \varepsilon$ whenever $m \geq N$ and $n \geq N$.

The converse of Exercise 10 is false. The sequence

$$\{2, 1.5, 1.42, 1.415, 1.4143, 1.41422, 1.414214, \ldots\}$$

given in Example 4(c) is Cauchy since terms past the kth term differ in, at worst, the kth decimal place, that is, by at most $1/10^k$. However, this sequence fails to converge in $\mathbb{Q}$. It converges to $\sqrt{2}$ in $\mathbb{R}$.

By Exercise 10, every sequence in $\mathbb{Q}$ or $\mathbb{R}$ that converges is Cauchy. By the previous example the converse is false in $\mathbb{Q}$. In the remainder of this section, we will prove that the converse is true in $\mathbb{R}$.

The sequence $\{n\}$ is not Cauchy. In fact, Cauchy sequences are never unbounded, as we show in the Exercise 13. First, we give some definitions.

Definition 12. A sequence $\{q_n\}$ in $\mathbb{Q}$ or in $\mathbb{R}$ is *bounded* if and only if there exists $q \in \mathbb{Q}$ such that $|q_n| \leq q$ for $n \in \mathbb{N}$. $\{q_n\}$ is *bounded above* if and only if there exists $q \in \mathbb{Q}$ such that $q_n \leq q$ for $n \in \mathbb{N}$. $\{q_n\}$ is *bounded below* if and only if there exists $q \in \mathbb{Q}$ such that $q_n \geq q$ for $n \in \mathbb{N}$.

A sequence is bounded if and only if it is bounded above and bounded below. On the other hand, a sequence can be bounded above and not bounded, or bounded below and not bounded as $\{-n\}$ and $\{n\}$ demonstrate.

> **Exercise 13.** Suppose $\{q_n\}$ is a Cauchy sequence in $\mathbb{Q}$ (or $\mathbb{R}$). Prove that $\{q_n\}$ is bounded.

Definition 14. A sequence $\{q_n\}$ in $\mathbb{Q}$ or in $\mathbb{R}$ is *increasing* if and only if $q_{n+1} \geq q_n$ for all $n \in \mathbb{N}$. $\{q_n\}$ is *decreasing* if and only if $q_{n+1} \leq q_n$ for all $n \in \mathbb{N}$. $\{q_n\}$ is *monotone* if and only if it is either increasing or decreasing.

A constant sequence $\{q\}$ is both increasing and decreasing! *Strictly* increasing and *strictly* decreasing sequence are defined using strict inequalities.

A subsequence is like a subset: a subsequence of a sequence $\{a_1, a_2, a_3, a_4, \ldots\}$ is a sequence
$$\{a_{n_1}, a_{n_2}, a_{n_3}, a_{n_4}, \ldots\}$$
where $n_1 < n_2 < n_3 < n_4 < \ldots$. The next theorem is usually called the Bolzano-Weierstrass Theorem.

Theorem 15. *Every bounded monotone sequence in $\mathbb{R}$ is convergent.*

Proof. Suppose $\{x_n\}$ is an increasing sequence in $\mathbb{R}$ that is bounded above. By the least upper bound property, the range of the sequence has a least upper bound x. We wish to show that the sequence converges to x. Since x is an upper bound, $x_n \leq x$ for all $n \in \mathbb{N}$. For $\varepsilon > 0$, there exists an $N \in \mathbb{N}$ such that $x - \varepsilon < x_N \leq x$, since otherwise $x - \varepsilon$ would be a smaller upper bound than the least upper bound x. Since the sequence is increasing,
$$x - \varepsilon < x_N \leq x_n \leq x$$
for $n \geq N$. This proves the convergence of $\{x_n\}$ to x.

The other case is left as an exercise. □

Exercise 16. Complete the proof of Theorem 15.

Lemma 17. *Every sequence in $\mathbb{R}$ has a monotone subsequence.*

Proof. Suppose $\{x_n\}$ is a sequence in $\mathbb{R}$. The set
$$P = \{n \in \mathbb{N} \mid x_n \geq x_m \text{ for all } m > n\}$$
is either infinite or finite. If you graph the sequence, the n's in P indicate "peaks" in the graph that are higher than all later points.

When P is infinite, write $P = \{n_1, n_2, n_3, n_4, \ldots\}$, where $n_{k+1} > n_k$ for all $k \in \mathbb{N}$. Then the subsequence
$$\{x_{n_1}, x_{n_2}, x_{n_3}, x_{n_4}, \ldots\}$$
is decreasing.

When P is finite, set n_1 greater than any $n \in P$, for example, $1 + \max(P)$. Since $n_1 \notin P$, there exists $n_2 > n_1$ such that $x_{n_1} < x_{n_2}$. Continue in this way to produce a subsequence that is strictly increasing. □

The result proved in the next exercise is also sometimes called the Bolzano-Weierstrass Theorem.

Exercise 18. Prove that every bounded sequence in $\mathbb{R}$ has a convergent subsequence.

We can now prove the converse of Exercise 10 in $\mathbb{R}$.

Theorem 19. *Every Cauchy sequence in $\mathbb{R}$ is convergent.*

13.3. Cantor's Set of Real Numbers

Proof. Suppose $\{x_n\}$ is a Cauchy sequence in $\mathbb{R}$. By Exercise 13, $\{x_n\}$ is bounded. By Lemma 17, it has a monotone subsequence $\{x_{n_k}\}$. By Theorem 15, $\{x_{n_k}\}$ converges; let x be the limit.

For $\varepsilon > 0$, since $\{x_n\}$ is Cauchy, there exists a natural number N such that $|x_m - x_n| < \frac{1}{2}\varepsilon$ whenever $m \geq N$ and $n \geq N$. Since $\{x_{n_k}\}$ converges to x, there exists a natural number M such that $|x_{n_k} - x| < \frac{1}{2}\varepsilon$ whenever $k \geq M$. Pick k such that $k \geq M$ and $n_k \geq N$. By the Triangle Inequality (Exercise 5), for $n \geq N$,

$$\begin{aligned} |x_n - x| &= |(x_n - x_{n_k}) + (x_{n_k} - x)| \\ &\leq |x_n - x_{n_k}| + |x_{n_k} - x| \\ &< \frac{1}{2}\varepsilon + \frac{1}{2}\varepsilon = \varepsilon. \end{aligned}$$

Therefore, the Cauchy sequence $\{x_n\}$ converges to x. □

13.3 Cantor's Set of Real Numbers

For Dedekind cuts, we defined a real number x as the set of all rational numbers less than x. Next we will define a real number x using Cauchy sequences.

Definition 20. Suppose that $x = \{x_n\}$ and $y = \{y_n\}$ are Cauchy sequences in $\mathbb{Q}$. Define a relation $\sim$ on the set of Cauchy sequences in $\mathbb{Q}$ by $x \sim y$ if and only if the sequence $\{x_n - y_n\}$ converges to 0.

That is, $x \sim y$ if and only if for every rational number $\varepsilon > 0$ there exists a natural number N such that $|x_n - y_n| < \varepsilon$ whenever $n \geq N$.

Exercise 21. Prove that the relation $\sim$ is an equivalence relation on the set of Cauchy sequences in $\mathbb{Q}$.

Definition 22. The set of equivalence classes of $\sim$ is the set of *Cantor's real numbers* that we denote by $\mathbb{R}_C$.

We wish to show that $\mathbb{R}$ and $\mathbb{R}_C$ are isomorphic, both being the set of all real numbers. We will define order and operations on $\mathbb{R}_C$.

It is natural to think that $[x] < [y]$ should mean $[y] - [x] > 0$ and hence $[y] - [x] > \varepsilon$ for some positive rational ε.

Definition 23. Define a relation $<$ on $\mathbb{R}_C$ for Cauchy sequences $x = \{x_n\}$ and $y = \{y_n\}$ of rational numbers by $[x] < [y]$ if and only if there exis a rational number $\varepsilon > 0$ and a natural number N such that $x_n + \varepsilon < y_n$ whenever $n \geq N$.

The definition of $[x] < [y]$ seems to depend on the choices of representatives of the equivalence classes $[x]$ and $[y]$. The next exercise shows that is not so.

Exercise 24. Prove that the relation $<$ on $\mathbb{R}_C$ is well-defined.

From here on we will assume that $\{x_n\}$ is a representative sequence for $[x]$ in $\mathbb{R}_C$.
We define $[x] \leq [y]$ to mean $[x] < [y]$ or $[x] = [y]$.

Example 25. Let $x_n = -1/n$ and $y_n = 1/n$ for $n \in \mathbb{N}$. Then $[x] = [y] = 0$ since $|x_n - y_n| = \frac{2}{n}$ converges to 0, and $x_n < y_n$ for all $n \in \mathbb{N}$. That is, there does not exist $N \in \mathbb{N}$ such that $y_n \leq x_n$ whenever $n \geq N$. Therefore, we could not define $\leq$ by comparing the terms x_n and y_n directly.

Exercise 26. Prove that $<$ is a (strict) total ordering on $\mathbb{R}_C$.

Next we define the operations.

Definition 27. Define *addition* on $\mathbb{R}_C$ by $[x] + [y] = [\{x_n + y_n\}]$.

Exercise 28. Prove that addition on $\mathbb{R}_C$ is well-defined.

Definition 29. Define *multiplication* on $\mathbb{R}_C$ by $[x][y] = [\{x_n y_n\}]$.

Exercise 30. Prove that multiplication on $\mathbb{R}_C$ is well-defined.

$\mathbb{R}_C$ completes $\mathbb{Q}$ in the way hinted at in Section 13.1. If $q \in \mathbb{Q}$, then it is represented by the constant sequence in $\mathbb{R}_C$, that is, $q = [\{q\}] \in \mathbb{R}_C$ is the embedded image of $q \in \mathbb{Q}$. So a sequence $\{x_n\}$ in $\mathbb{Q}$ can be viewed as a sequence $\{x_n\}$ in $\mathbb{R}_C$.

Exercise 31. If $x = \{x_n\}$ is a Cauchy sequence in $\mathbb{Q}$, then $\{x_n\}$ converges to $[x]$ in $\mathbb{R}_C$.

Next we could check that $\mathbb{R}_C$ with this ordering and these operations has all the properties of an ordered field with the least upper bound property and that $\mathbb{Q}$ is embedded as a subfield. Only the least upper bound property would give us any difficulty. We know all such structures are isomorphic and, in particular, $\mathbb{R}_C$ is isomorphic to $\mathbb{R}$.

However, we will construct the isomorphism directly.

13.4 The Isomorphism from Cantor's to Dedekind's Reals

We next give an isomorphism $f : \mathbb{R} \to \mathbb{R}_C$. Take a cut representing x in $\mathbb{R}$. If we can extract a sequence from the cut that converges to x, this gives us a way to define a candidate for our isomorphism. It is harder to define the isomorphism in the opposite direction. This makes sense since we know all about $\mathbb{R}$ but we know almost nothing about $\mathbb{R}_C$. We will make frequent use of the properties of $\mathbb{R}$ in proving the existence of the isomorphism.

The cut x is the set

$$\{q \in \mathbb{Q} \mid q < x \text{ in } \mathbb{R}\}.$$

We will construct a sequence of rationals in x. First, pick

$$q_1 \in x.$$

Suppose that we have chosen $q_1, \cdots, q_n$ (in a manner we will describe). If $q_{n+1} + 1 < x$, let

$$q_{n+1} = q_n + 1.$$

13.4. The Isomorphism from Cantor's to Dedekind's Reals

Otherwise, if $q_{n+1} + \frac{1}{2} < x < q_{n+1} + 1$, let

$$q_{n+1} = q_n + \frac{1}{2^k}$$

such that

$$q_{n+1} + \frac{1}{2^k} < x < q_{n+1} + \frac{1}{2^{k-1}}.$$

This defines an increasing sequence $\{q_n\}$ in $\mathbb{Q}$. We will check that it is Cauchy and then define $f : \mathbb{R} \to \mathbb{R}_C$ by

$$f(x) = [\{q_n\}].$$

Exercise 32. Prove that the process in the last paragraph defines a sequence and that it is Cauchy in $\mathbb{Q}$ and converges to x in $\mathbb{R}$.

We define $f : \mathbb{R} \to \mathbb{R}_C$ by $f(x) = [\{q_n\}]$. We must check that this is well defined, that is, independent of the choice of q_1.

Exercise 33. Prove that f is well defined.

It remains to show that f is an isomorphism, that is, a structure-preserving bijection.

Exercise 34. Prove that f is order preserving and hence one-to-one.

Exercise 35. Prove that f is addition preserving.

Exercise 36. Prove that f is multiplication preserving.

Exercise 37. Prove that f is onto.

This completes the proof that the structure $\mathbb{R}_C$ is the same as the set of real numbers $\mathbb{R}$ with its order relation and operations.

13.5 Supplemental Exercises

Definition Review. There were fifteen italicized terms defined in this chapter. Define each of the following and give an example of each:

Definition 1. a *sequence* in a set, X

Definition 3. a sequence $\{q_n\}$ in $\mathbb{Q}$ or $\mathbb{R}$) *converges* to q

Definition 3. a sequence $\{q_n\}$ in $\mathbb{Q}$ or $\mathbb{R}$) *convergent*

Definition 11. a sequence $\{q_n\}$ in $\mathbb{Q}$ or $\mathbb{R}$ is *Cauchy* sequence

Definition 12. a sequence $\{q_n\}$ in $\mathbb{Q}$ or $\mathbb{R}$ is *bounded*

Definition 12. a sequence $\{q_n\}$ in $\mathbb{Q}$ or $\mathbb{R}$ is *bounded above*

Definition 12. a sequence $\{q_n\}$ in $\mathbb{Q}$ or $\mathbb{R}$ is *bounded below*

Definition 14. a sequence $\{q_n\}$ in $\mathbb{Q}$ or $\mathbb{R}$ is *increasing*

Definition 14. a sequence $\{q_n\}$ in $\mathbb{Q}$ or $\mathbb{R}$ is *decreasing*

Definition 14. a sequence $\{q_n\}$ in $\mathbb{Q}$ or $\mathbb{R}$ is *monotone*

Definition 20. a relation $\sim$ on the set of all Cauchy sequences in $\mathbb{Q}$ (that will define Cantor's real numbers)

Definition 22. the set, $\mathbb{R}_C$, of *Cantor's real numbers*

Definition 23. the order relation $<$ on $\mathbb{R}_C$

Definition 27. *addition* in $\mathbb{R}_C$

Definition 29. *multiplication* in $\mathbb{R}_C$

Supplemental Exercise 2. Project. Revisit the Project in Supplemental Exercise 1 from Chapter 12: Suppose we say that a real number is a number of the form

$$a.a_1 a_2 a_3 a_4 \ldots .$$

This may involve infinite sums. How would you define addition and multiplication of real numbers so defined?

14

The Complex Numbers

14.1 Introduction

In this chapter, we will give a rigorous construction of the field $\mathbb{C}$ of complex numbers. The addition and multiplication operations on $\mathbb{C}$ will be defined to preserve the operations on $\mathbb{R}$. We will show that $\mathbb{C}$ is a field containing $\mathbb{R}$ as a subfield. Let us start informally.

Complex numbers arise when we desire solutions of equations like $x^2 = -1$. We declare that there is a solution i. Having defined i, we allow the standard algebraic operations on the real numbers to work on (the closure of) the set containing the real numbers together with the new number i, which has the property that $i^2 = -1$. The complex numbers defined in this way are of the form $a + bi$, where a and b are real numbers. We need to define the multiplication bi and the addition $a + bi$.

For a moment, let us suppose that we have defined the complex numbers. Using familiar algebraic properties, we see that

$$(-i)^2 = (-1)^2(i)^2 = (1)(-1) = -1.$$

So $x^2 = -1$ has two solutions! This agrees with the notion that quadratic polynomials have two roots. The choice in the definition of i appears in Definition 14 below. However, this choice is inconsequential as Exercise 20 will show.

While addition and multiplication of complex numbers is natural, ordering the complex numbers is not. Suppose we have a total ordering on $\mathbb{C}$ that agrees with the usual ordering on the subset $\mathbb{R}$. Since a total ordering is trichotomous, one of the following holds: $i = 0$, $i > 0$, $i < 0$. $i = 0$ is absurd. Suppose that $i > 0$. If $\mathbb{C}$ is to be an ordered field, then multiplying positives yields a positive. That is, $i^2 > 0$. But $-1 \not> 0$ in $\mathbb{R}$. The other case is no better, as you can check. This will be made precise in Exercises 17 and 18.

14.2 Algebra of Complex Numbers

We first define the set $\mathbb{C}$ of complex numbers as well as the operations of addition and multiplication on $\mathbb{C}$.

Definition 1. The set of *complex* numbers is the set $\mathbb{C} = \mathbb{R}^2$.

You may have seen a complex number defined to be one of the form $a + bi$ where a and b are real numbers. We can equate an ordered pair (a, b) from Definition 1 with $a + bi$.

By formally applying what algebra suggests, we can define addition by

$$(a + bi) + (c + di) = (a + c) + (b + d)i$$

and multiplication by

$$(a + bi)(c + di) = (ac - bd) + (ad + bc)i.$$

This yields the following definitions.

Definition 2. *Addition* of complex numbers (a, b) and (c, d) is defined by

$$(a, b) + (c, d) = (a + c, b + d).$$

Definition 3. *Multiplication* of complex numbers (a, b) and (c, d) is defined by

$$(a, b)(c, d) = (ac - bd, ad + bc).$$

These operations are well-defined; unlike the constructions of the integers and the rationals, this construction does not use equivalence classes. We next verify that $\mathbb{C}$ is a field.

We start with the properties of addition.

Exercise 4. Prove that addition on $\mathbb{C}$ is commutative.

Exercise 5. Prove that addition on $\mathbb{C}$ is associative.

Exercise 6. Prove that there exists a unique additive identity for $\mathbb{C}$.

We let 0 denote the additive identity; that is, $0 = (0, 0)$.

Exercise 7. Prove that every element of $\mathbb{C}$ has a unique additive inverse.

Denote the additive inverse of a number $z \in \mathbb{C}$ by $-z$.
Now, we get the properties of multiplication.

Exercise 8. Prove that multiplication on $\mathbb{C}$ is commutative.

Exercise 9. Prove that multiplication on $\mathbb{C}$ is associative.

Exercise 10. Prove that there exists a unique multiplicative identity 1 for $\mathbb{C}$.

We let 1 denote the multiplicative identity $(1, 0)$.

Exercise 11. Prove that every nonzero $z \in \mathbb{C}$ has a unique multiplicative inverse.

Denote the multiplicative inverse of a number $z \in \mathbb{C}$ by z^{-1}.
It remains to prove the distributive property.

14.3. Order on the Complex Field

> **Exercise 12.** Prove the distributive property for $\mathbb{C}$.

This completes the proof that $\mathbb{C}$ is a field. We will come back to the question of order on $\mathbb{C}$ and whether $\mathbb{C}$ is an ordered field in the next section.

Since the product of real numbers

$$(a, 0)(b, 0) = (ab, 0)$$

looks a lot like the equation $a \cdot b = ab$, we will write a for $(a, 0)$ and b for $(b, 0)$. This suggests a candidate for an embedding of $\mathbb{R}$ in $\mathbb{C}$, which is postponed until Exercise 19.

Next, we will identify the imaginary number i.

> **Exercise 13.** Show that $(0, 1)^2 = -1$. Is there any other complex number whose square equals -1?

Definition 14. The element $(0, 1) \in \mathbb{C}$ is called i.

We have made a choice in Definition 14 since $(0, -1)^2 = -1$ also. The nature of this choice will be explained by Exercise 20.

Since

$$\begin{aligned}(a, b) &= (a, 0) + (0, b) = (a, 0)(1, 0) + (b, 0)(0, 1) \\ &= (a, 0) \cdot 1 + (b, 0) \cdot i = a \cdot 1 + b \cdot i,\end{aligned}$$

we will write $a + bi$ for $(a, b) \in \mathbb{C}$. We call a the *real part* of (a, b) and b the *imaginary part* of (a, b). For $b \in \mathbb{R}$, the numbers $(0, b) \in \mathbb{C}$ are called *imaginary*.

14.3 Order on the Complex Field

Let us look at order relations on $\mathbb{C}$. Whenever we take products of ordered sets, we can define an order relation on the product called the *lexicographic* or *dictionary* order. We will do this for $\mathbb{C}$.

Definition 15. Define the *lexicographic ordering* $\prec$ by $(a, b) \prec (c, d)$ if and only if $a < c$, or $a = c$ and $b < d$.

> **Exercise 16.** Prove that $\prec$ is a total ordering on $\mathbb{C}$ and that $(a, 0) \prec (b, 0)$ if and only if $a < b$.

> **Exercise 17.** Prove that $\prec$ does not make $\mathbb{C}$ an ordered field.

The lexicographic ordering is not a bad choice since no other ordering makes $\mathbb{C}$ an ordered field either. To prove this consider both of the following possibilities: $-1 < 0$ or $-1 > 0$.

> **Exercise 18.** Prove that there is no total ordering on $\mathbb{C}$ that makes it an ordered field.

14.4 Embedding the Reals in the Complex Numbers

$\mathbb{R}$ is not a subset of $\mathbb{C}$ as we have defined $\mathbb{C}$. We can find an embedding of $\mathbb{R}$ into $\mathbb{C}$ that preserves the operations so that we may think of $\mathbb{R}$ as a subfield of $\mathbb{C}$.

> **Exercise 19.** Prove that there exists an embedding of $\mathbb{R}$ into $\mathbb{C}$.

Perhaps we were too terse in Exercise 19; let us elaborate. So far, our embeddings have had the property of preserving order as well as operations. If you thought of $\mathbb{C}$ with the lexicographic ordering, then the ordering on $\mathbb{R}$ is compatible. However, since $\mathbb{C}$ does not have a natural ordering (i.e., one that makes it an ordered field), we have in mind a weaker kind of embedding that does not include the order-preserving property, an embedding of a field rather than an ordered field.

We would be done with complex numbers now except for one small detail. The definition of i left us a choice as the following exercise explains.

> **Exercise 20.** Prove that there are exactly two automorphisms $f : \mathbb{C} \to \mathbb{C}$ such that $f(x) = x$ for all $x \in \mathbb{R}$.

Since the automorphisms in Exercise 20 induce the identity $I_\mathbb{R}$ when restricted to $\mathbb{R}$, we say that these automorphisms of $\mathbb{C}$ *extend* the identity on $\mathbb{R}$.

Suppose $f : \mathbb{C} \to \mathbb{C}$ is an automorphism such that $f(\mathbb{R}) = \mathbb{R}$. By Exercise 41 of Chapter 12, $I_\mathbb{R}$ is the unique automorphism on $\mathbb{R}$. The induced function on $\mathbb{R}$ preserves operations and is one-to-one since f has these properties, making it an isomorphism onto its range $\mathbb{R}$; the induced function is an automorphism of $\mathbb{R}$ and hence equal to $I_\mathbb{R}$. Therefore, the statement in Exercise 20 can be strengthened: there are exactly two automorphisms $f : \mathbb{C} \to \mathbb{C}$ such that $f(\mathbb{R}) = \mathbb{R}$.

14.5 Supplemental Exercises

Definition Review. There were five definitions in this chapter. Define each of the following and give an example of each:

Definition 1. the set, $\mathbb{C}$, of *complex* numbers

Definition 2. *addition* in $\mathbb{C}$

Definition 3. *multiplication* in $\mathbb{C}$

Definition 14. i

Definition 15. the *lexicographic ordering* on $\mathbb{C}$

Supplemental Exercise 1. Prove that, for $q, z \in \mathbb{C}$, $wz = 0$ if and only if $w = 0$ or $z = 0$.

Supplemental Exercise 2. Project. Investigate the history of another number system such as the quaternions, surreal numbers, or hyperreal numbers. Describe their construction and utility.

Supplemental Exercise 3. Project. In an 1825 letter ([5, page 112]), Gauss wrote to Hansen about the imaginary number i and said: "Those investigations penetrate deeply into many others, I may even say into the metaphysics of the theory of space, and only with difficulty can I tear myself away from such results arising therefrom, as, for example, the true metaphysics of negative and imaginary quantities. The true meaning of i stands very vividly before my soul, but it will be very difficult to put it into words, which can only give but a vague fleeting image." Investigate the history of the acceptance of complex numbers by mathematicians.

IV

TIME SCALES

15

Time Scales

15.1 Introduction

In Chapter 5, we defined one of the most important concepts in mathematics, the function. You proved many results concerning the domain and the range. We defined some important characteristics of functions such as injectivity, surjectivity, and bijectivity. Also, we discussed composite functions and what is required for a function to have an inverse.

Most of your previous experience has probably been with functions defined on subsets of the real line such as intervals or the whole real line. Perhaps, you have some experience with functions, such as sequences, that are defined on a discrete set such as the integers or a subset of it.

Let's join these two concepts and expand our view to consider functions that are defined on sets that includes both of these kinds of sets. They are known in the literature as a *time scales*. Stefan Hilger developed the calculus on time scales in 1988. Time scales are the basis for the development of an analytical structure that unifed the theory of differential equations and their discrete version, known as difference equations. (You do not need to know what these words mean!)

15.2 Preliminary Results

We begin with the definition of a sequence of real numbers, as in Definition 1 in Chapter 13.)

Definition 1. A *sequence* is a function, $f : \mathbb{N} \to \mathbb{R}$, from the natural numbers to the reals.

We denote a general sequence by $\{a_n\}$ where $a_n = f(n)$ for all $n \in \mathbb{N}$. We can generalize Definition 1: a sequence $\{a_n\}$ is *in a set S* if and only if $a_n \in S$ for all $n \in \mathbb{N}$.

We can think of the range of f as a collection of real numbers with an ordering, that is, we have a first term, a second term, etc.

Remark 2. The function f defining a sequence need not be one-to-one. For example, the constant sequence given by $f(n) = 0$ for all $n \in \mathbb{N}$. The range of this function is the singleton set $\{0\}$. The sequence may be denoted by

$$\{0, 0, 0, 0, 0, \ldots\}.$$

To make things slightly more confusing, we can denote this constant sequence by $\{0\}$.

Context will let us know when we are thinking of a sequence.

Example 3. Here are two more sequences.
(a) If $f(n) = 2n$, then $\{a_n\} = \{2, 4, 6, 8, \ldots\}$ and we have the set of even natural numbers.
(b) In many cases f is defined by describing the first few terms of the sequence, as

$$\{a_n\} = \{1, -1, 2, -2, 3, -3, \ldots\}.$$

From this brief list you should be able to determine the following terms.

As n increases, the terms of the sequence may approach a particular value. We call this value the limit of the sequence. For example, the sequence

$$\left\{1, \frac{1}{2}, \frac{1}{4}, \frac{1}{16}, \ldots\right\}$$

seems to "approach" the number 0. In the language of mathematics, we have a precise way to describe what it means to "approach."

Definition 4. The sequence $\{a_n\}$ has a *limit* L if and only if for every $\varepsilon > 0$ there exists $N \in \mathbb{N}$ such that if $n \geq N$, then

$$|a_n - L| < \varepsilon.$$

If the limit L exists, we say that $\{a_n\}$ *converges* to L and denote it by $\{a_n\} \to L$.

What does the inequality in the definition tell us about how the terms of the sequence "approach" the limit L? We know that

$$|a_n - L| < \varepsilon \quad \text{if and only if} \quad L - \varepsilon < a_n < L + \varepsilon.$$

For convergence, this must hold for all $n \geq N$. That is, for all natural numbers n that are larger than N, a_n stays within ε of L. In addition, this must hold for any positive $\varepsilon > 0$. When we choose a smaller ε we may have to find a new N. If the limit exists, we know that this can always be done.

Now we use the concept of limit to define a closed subset of the reals.

Definition 5. A subset of the reals is *closed* if and only if all limits of all convergent sequences in the set are contained in the set.

That is, Definition 5 says that a subset S of $\mathbb{R}$ is closed when every sequence of elements in S that converges in $\mathbb{R}$ actually converges in S. Notice that $\mathbb{R}$ and $\emptyset$ are closed. (See Supplemental Exercises 1 and 2.)

Is an interval that contains its endpoints closed? After all, we call this a closed interval. Let's use Definition 5 to verify that this is true.

Theorem 6. *Let $a, b \in \mathbb{R}$ with $a < b$. The closed interval $[a, b]$ is a closed set.*

Proof. We need to verify that the limit of every convergent sequence in $[a, b]$ is in the set $[a, b]$. Let's proceed by contradiction. Assume that there exists a sequence $\{a_n\}$ in $[a, b]$ and a real number $L \notin [a, b]$ such that $\{a_n\}$ converges to L. There are two cases: $L < a$ or $b < L$.

15.3. The Time Scale and its Jump Operators

Case 1: Suppose that $L < a$. From Theorem 30 of Chapter 8, we can choose $\varepsilon > 0$ such that
$$\varepsilon < \frac{a-L}{2}.$$

Then, from Definition 4, there exists $N_0 \in \mathbb{N}$ such that
$$|a_n - L| < \varepsilon < \frac{a-L}{2}$$

for $n \geq N_0$. Therefore,
$$a_n - L < \frac{a-L}{2}$$

or, equivalently,
$$a_n < L + \frac{a-L}{2} = \frac{a+L}{2} < \frac{a+a}{2} = a$$

since $L < a$. Hence, $a_n \notin [a,b]$ for $n \geq N_0$. This contradicts our assumption $a_n \in [a,b]$ for all $n \in \mathbb{N}$.

Case 2: The proof of this case is analogous to Case 1. □

Exercise 7. Complete the proof of Theorem 6 by proving Case 2.

Proposition 8. *The interval $[a,b)$ is not closed.*

Proof. We need to show that there exists a sequence, $\{a_n\}$ in $[a,b)$, with a limit $L \notin [a,b)$. For this purpose, let
$$a_n = a + \left(1 - \frac{1}{2^n}\right)(b-a)$$
$$= b - \frac{1}{2^n}(b-a).$$

As $n \to \infty$, $\frac{1}{2^n}(b-a) \to 0$. We can see this from the definition of convergence because for every $\varepsilon > 0$ there exists $N \in \mathbb{N}$ such that $\frac{1}{2^n}(b-a) < \varepsilon$ for $n \geq N$. Therefore $\{a_n\} \to b \notin [a,b)$. □

Exercise 9. Prove that the interval $(a,b]$ is not closed.

15.3 The Time Scale and its Jump Operators

We are now ready to define time scales. It is straightforward.

Definition 10. A *time scale* is a nonempty closed subset of the reals, $\mathbb{R}$.

We frequently use the notation $\mathbb{T}$ to denote a time scale. Let's look at some examples.

Example 11. $\mathbb{R}$ and $\mathbb{Z}$ are time scales.

Example 12. A finite union of nonempty closed sets is closed and hence a time scale. For example,
$$\mathbb{T} = \bigcup_{j=1}^{5} \left[j, j + \frac{1}{2}\right]$$
is a time scale.

Example 13. What happens with an infinite union of closed sets? If the union is of finitely many closed sets, as in Example 12 above, we certainly have a time scale. In the infinite union
$$\mathbb{T} = \bigcup_{j=2}^{\infty} \left[\frac{1}{j}, 1\right],$$
as $j \to \infty$ the intervals get larger. The left endpoint of the intervals converges to zero, but is never equal to zero. Therefore,
$$\mathbb{T} = \bigcup_{j=2}^{\infty} \left[\frac{1}{j}, 1\right] = (0, 1].$$
$\mathbb{T}$ is not a closed set and not a time scale.

Example 14. Let's look at nonempty intersections of closed sets as well. If
$$\mathbb{T} = \bigcap_{j=1}^{5} \left[\frac{1}{j+1}, 1\right],$$
then the interval for $j = 1$, $[\frac{1}{2}, 1]$, is contained in all the other intervals. Hence $\mathbb{T} = [\frac{1}{2}, 1]$ is a time scale.

> **Exercise 15.** Prove that the infinite intersection of closed sets is either empty or a time scale.

Inherent to time scales is the possibility of gaps between points or closed intervals within the set. A set of isolated points is a time scale if it is closed. So, $\mathbb{T} = \{1, 2, 3, 5\}$ is a time scale. If
$$\mathbb{S} = \left\{\frac{1}{n} \,\middle|\, n \in \mathbb{N}\right\},$$
it is an infinite set of isolated points but is it closed? The set is formed by a sequence with limit 0. To create a time scale from this set we must include the limit, 0.

In the next section, we say what it means for a function, $f : \mathbb{T} \to \mathbb{R}$, to have a derivative. This will require that we find some way to move through the gaps that are possible in its domain, $\mathbb{T}$. For this purpose, we have forward and backward jump operators:

Definition 16. For a time scale $\mathbb{T}$,
(a) the *forward jump operator* on $\mathbb{T}$, $\sigma : \mathbb{T} \to \mathbb{T}$, is the function defined by
$$\sigma(t) = \inf\{s \in \mathbb{T} \mid s > t\},$$
where $\inf \emptyset = \sup \mathbb{T}$, and
(b) the *backward jump operator* on $\mathbb{T}$, $\rho : \mathbb{T} \to \mathbb{T}$, is the function defined by
$$\rho(t) = \sup\{s \in \mathbb{T} \mid s < t\},$$
where $\sup \emptyset = \inf \mathbb{T}$.

15.4. Limits and Continuity

Recall from Definition 25 of Chapter 6 that the supremum of a set, denoted by sup, is the least upper bound of the set. In the same chapter, we introduced the analogous concept, greatest lower bound, known as the infimum of a set and denoted by inf. These are not binary operators as defined in Chapter 8.

What do the forward and backward jump operators tell us about the element $t \in \mathbb{T}$ that they operate on? In the case of the forward jump operator, $\sigma(t)$ is the succeeding element in the time scale, if such an element exists. Let's look at some examples.

Example 17. Define a time scale $\mathbb{T} = [2, 5] \cup \{7, 8.5, 11\}$. From Definition 16, and the preceding comment, $\sigma(5) = 7$, $\sigma(7) = 8.5$, and $\sigma(8.5) = 11$.

What is the value of $\sigma(3)$? Because σ takes $\mathbb{T}$ to $\mathbb{T}$, the range element must be in the time scale. So, as Definition 16 describes, the least upper bound of the set of elements $s \in \mathbb{T}$ such that $s > 3$. It is 3 itself.

What is the value of $\sigma(11)$? Since $\{s \in \mathbb{T} \mid s > 11\} = \emptyset$, by Definition 16, $\sigma(11) = \inf \emptyset = \sup \mathbb{T} = 11$.

The backward jump operator gives us analogous information, but in the opposite direction.

Exercise 18. For the time scale $\mathbb{T}$ defined in Example 17, calculate the values of $\rho(5)$, $\rho(7)$, $\rho(11)$, and $\rho(2)$.

We classify points in a time scale using the jump operators.

Definition 19. Let $t \in \mathbb{T}$. Then t is called
  right scattered if and only if $t < \sigma(t)$,
  right dense if and only if $t = \sigma(t)$,
  left scattered if and only if $\rho(t) < t$, and
  left dense if and only if $\rho(t) = t$.
If $\rho(t) < t < \sigma(t)$, the point t is called *isolated*. If $\rho(t) = t = \sigma(t)$, t is is called *dense*.

Remark 20. Maximal and minimal elements of a time scale can be treated as "one-sided." For example, if $t = \sup(\mathbb{T})$ and t is left scattered, we think of t as being isolated even though $\rho(t) < t = \sigma(t) = \sup(\mathbb{T})$.

Interesting situations arise when a point is right dense and left scattered, or left dense and right scattered.

15.4 Limits and Continuity

We will define the limit of a function and use this to define what it means for a function to be continuous on $\mathbb{T}$.

Remark 21. When we are using an interval of the reals in our discussions, we mean the interval intersected with the time scale. The notation we use for this is

$$[a, b]_{\mathbb{T}} = \{x \in \mathbb{T} \mid a \leq x \leq b\} = \mathbb{T}\widehat{[a, b]}.$$

Definition 22. A function $f : \mathbb{T} \to \mathbb{R}$ has a *limit* L at $t_0 \in \mathbb{T}$ if and only if for every $\varepsilon > 0$ there exists $\delta > 0$ such that if $t \in [t_0 - \delta, t_0 + \delta]_{\mathbb{T}}$ then

$$|f(t) - L| < \varepsilon.$$

If t_0 is an isolated point, then $L = f(t_0)$.

If the limit exists, we write $\lim_{t \to t_0} f(t) = L$.

Remark 23. If t_0 is an isolated point, the definition is vacuously true. This is equivalent to saying that limits always exist at isolated points in the time scale. Also, if t_0 is a dense point then the definition of limit is similar to the one you would see in your calculus course.

Remark 24. For points t_0 that are dense on one side and scattered on the other, determining if the limit of a function at t_0 exists is similar to determining if a one-sided limit exists. When δ is small enough, the only elements $t \in \mathbb{T}$ within δ of t_0 are on the dense side of t_0.

Now we use the definition of a limit to define a continuous function $f : \mathbb{T} \to \mathbb{R}$.

Definition 25. A function $f : \mathbb{T} \to \mathbb{R}$ is *continuous* at a point $t_0 \in \mathbb{T}$ if and only if

$$\lim_{t \to t_0} f(t) = f(t_0).$$

Remark 26. As with the limit of a function at dense points, continuity of a function at dense points is essentially the same as continuity of a function on a subset of the real line. For points that are dense on one side and scattered on the other, the limit required by the definition is essentially a one sided limit.

Exercise 27. Prove that a function $f : \mathbb{T} \to \mathbb{R}$ is continuous at an isolated point in $\mathbb{T}$.

We define what it means for a function to be continuous on its domain $\mathbb{T}$ in a natural way.

Definition 28. A function $f : \mathbb{T} \to \mathbb{R}$ is *continuous* if and only if it is continuous at every $t \in \mathbb{T}$.

Continuity is a statement about what is happening at dense points (functions are always continuous at isolated points). So continuity and discontinuity described by the examples below should have a familiar ring.

Let's start with some basic examples.

Example 29. Let $f : \mathbb{T} \to \mathbb{R}$ be a constant function, $f(t) = C$. Let's determine if f is continuous on every time scale, $\mathbb{T}$. Let $\varepsilon > 0$ and $t_0 \in \mathbb{T}$ be given. Then, for $t \in \mathbb{T}$,

$$|f(t) - f(t_0)| = |C - C| = 0 < \varepsilon.$$

From Definition 28, we know that f is continuous at every $t_0 \in \mathbb{T}$. In this example, $|f(t) - f(t_0)| = 0$ is always satisfied, independent of δ. So we can choose δ however we please. It is common to choose $\delta = \varepsilon$ or $\delta = 1$.

15.4. Limits and Continuity

Example 30. Let $f : \mathbb{T} \to \mathbb{T}$ be the identity function, $f(t) = t$. Let $\varepsilon > 0$ be given and suppose $|t - t_0| < \delta$ for some $\delta > 0$. Then we have

$$|f(t) - f(t_0)| = |t - t_0| < \delta.$$

If we define $\delta = \varepsilon$ we have accomplished our goal.

Example 31. Now let's look at a given time scale and a more interesting function. In the time scale

$$\mathbb{T} = \{2\} \cup \left\{ 2 + \frac{1}{n} \mid n \in \mathbb{N} \right\}$$

all points are isolated except for the dense point 2. Define $f : \mathbb{T} \to \mathbb{R}$ by

$$f(t) = \begin{cases} \frac{1}{t-2.3}, & t \neq 2 + \frac{1}{3} \\ 11, & t = 2 + \frac{1}{3}. \end{cases}$$

What can we say about continuity on the domain $\mathbb{T}$? At $2 + \frac{1}{3}$, although the function jumps to 11 at this point, the point is isolated and therefore f is continuous at this and its other isolated points.

At 2, we have a dense point so we need to look at the limit at this point (from the right because it is the minimal point of $\mathbb{T}$). Let $L = f(2) = \frac{1}{-.3}$ and $\varepsilon > 0$ be given. Suppose that $|t - 2| < \delta$ for some $\delta > 0$ and $t \in \mathbb{T}$. Then

$$|f(t) - f(2)| = \left| \frac{1}{t-2.3} + \frac{1}{.3} \right| = \left| \frac{t-2}{.3(t-2.3)} \right|, \tag{1}$$

if $\delta < 1/3$. Since $|t - 2| < \delta$,

$$\left| \frac{t-2}{.3(t-2.3)} \right| < \frac{\delta}{.3|t-2.3|}. \tag{2}$$

Using the inequality $|x - y| \geq |x| - |y|$, which holds for all real numbers x and y (see Supplemental Exercise 10), and the inequality $-|t - 2| > -\delta$, we get

$$|t - 2.3| = |.3 - (t - 2)|$$
$$\geq .3 - |t - 2|$$
$$> .3 - \delta.$$

Therefore,

$$\frac{1}{|t - 2.3|} < \frac{1}{.3 - \delta},$$

which we substitute back into (2) to get

$$\frac{\delta}{.3|t - 2.3|} < \frac{\delta}{.3(.3 - \delta)}, \tag{3}$$

whenever $\delta < .3 < 1/3$. By (1), (2), and (3), if $\delta < .3$, then

$$|f(t) - f(2)| < \frac{\delta}{.3(.3 - \delta)}.$$

If we set the last expression equal to ε, then we have

$$\frac{\delta}{.3(.3-\delta)} = \varepsilon \iff \delta = \varepsilon(.3)(.3-\delta)$$
$$\iff \delta - .3\varepsilon\delta = .09\varepsilon$$
$$\iff \delta(1 - .3\varepsilon) = .09\varepsilon$$
$$\iff \delta = \frac{.09\varepsilon}{1 - .3\varepsilon},$$

whenever $\varepsilon < 10/3$. So, given $\varepsilon > 0$, we can set δ equal to the lesser of $.3$ and

$$\frac{.09\varepsilon'}{1-.3\varepsilon'},$$

where ε' is the lesser of ε and $10/3$. This is the desired δ that proves the continuity of f at 2.

Remark 32. The moral of Example 31 is that computations are difficult or, at least, take a lot of time. This was a "quick and easy" computation!

We can eliminate the necessity of determining δ as a function of ε, which can be difficult if not impossible, by using the following equivalent theorem to prove continuity.

Theorem 33. Let $g : \mathbb{R}_+ \to \mathbb{R}_+$ with $g(0) = 0$ and $g(\delta) \to 0$ as $\delta \to 0$. Let $f : \mathbb{T} \to \mathbb{R}$ and $t_0 \in \mathbb{T}$. f is continuous at t_0 if and only if for every $\varepsilon > 0$ there exists $\delta > 0$ such that if $t \in [t_0 - \delta, t_0 + \delta]_\mathbb{T}$ then

$$|f(t) - f(t_0)| < g(\varepsilon).$$

In either case, $\lim_{t \to t_0} f(t) = f(t_0)$.

Proof. Suppose $f : \mathbb{T} \to \mathbb{R}$ and

$$|f(t) - f(t_0)| < g(\delta)$$

when $|t - t_0| < \delta$. Choose $\varepsilon > 0$. From the nature of g, we know there exists $\delta_0 > 0$ such that for $\delta < \delta_0$, $|g(\delta)| < \varepsilon$. So, for $|t - t_0| < \delta < \delta_0$, we have

$$|f(t) - f(t_0)| < g(\delta) < \varepsilon.$$

Therefore, by definition, f is continuous at t_0. $\square$

> **Exercise 34.** Explain how Theorem 33 can be used in Example 31. Can we avoid solving for δ in terms of ε?

Another way to characterize continuity incorporates the nature of the domain. Let $\mathbb{T}$ be a time scale. The continuity of a function $f : \mathbb{T} \to \mathbb{R}$ is equivalent to the inverse image of every time scale—moreover, every closed subset of $\mathbb{T}$—is either a time scale or the empty set. See Supplemental Exercise 18.

15.5 Supplemental Exercises

Definition Review. There were thirteen italicized terms defined in this chapter. Define each of the following and give an example of each:

Definition 1. *sequence*

Definition 4. *limit* of a sequence

Definition 5. a *closed* subset of the reals

Definition 10. *time scale*

Definition 16. *forward jump operator*

Definition 16. *backward jump operator*

Definition 19. *right-scattered* point

Definition 19. *right-dense* point

Definition 19. *left-scattered* point

Definition 19. *left-dense* point

Definition 22. the *limit* L of a function $f : \mathbb{T} \to \mathbb{R}$ at $t_0 \in \mathbb{T}$

Definition 25. a function $f : \mathbb{T} \to \mathbb{R}$ is *continuous* at a point $t_0 \in \mathbb{T}$

Definition 28. a function $f : \mathbb{T} \to \mathbb{R}$ is *continuous*

Supplemental Exercise 1. Show that $\mathbb{R}$ is a closed subset of $\mathbb{R}$.

Supplemental Exercise 2. Show that $\emptyset$ is a closed subset of $\mathbb{R}$.

Supplemental Exercise 3. By translating the algebraic construction in the proof of Proposition 8 into geometrical language, can you see how the sequence in the proof was devised?

Supplemental Exercise 4. Show that the limit of a sequence, if it exists, is unique.

Supplemental Exercise 5. Create a time scale that contains

$$\left\{ \frac{1}{n} \, \bigg| \, n \in \mathbb{N} \right\} \cup \left\{ 1 - \frac{1}{n} \, \bigg| \, n \in \mathbb{N} \right\}.$$

What is the smallest such time scale?

Supplemental Exercise 6. Show that the forward and backwards jump operators are not necessarily inverse functions.

Supplemental Exercise 7. Show that a point in a time scale cannot be both left scattered and left dense.

Supplemental Exercise 8. Prove that $f : \mathbb{T} \to \mathbb{R}$, $f(t) = t^2$ is continuous on $\mathbb{T}$.

Supplemental Exercise 9. Prove that $|x + y| \leq |x| + |y|$ for all real numbers x and y.

Supplemental Exercise 10. Prove that $|x - y| \geq |x| - |y|$ for all real numbers x and y.

Supplemental Exercise 11. Graph the function f in Example 31.

Supplemental Exercise 12. Let $f : \mathbb{T} \to \mathbb{R}$ and $g : \mathbb{T} \to \mathbb{R}$ be continuous functions on $\mathbb{T}$. Prove that $f + g$ is continuous on $\mathbb{T}$.

Supplemental Exercise 13. Let $f : \mathbb{T} \to \mathbb{R}$ and $g : \mathbb{T} \to \mathbb{R}$ be continuous functions on $\mathbb{T}$. Prove that $f - g$ is continuous on $\mathbb{T}$.

Supplemental Exercise 14. Let $f : \mathbb{T} \to \mathbb{R}$ and $g : \mathbb{T} \to \mathbb{R}$ be continuous functions on $\mathbb{T}$. Prove that fg is continuous on $\mathbb{T}$.

Supplemental Exercise 15. Let $f : \mathbb{T} \to \mathbb{R}$ and $g : \mathbb{T} \to \mathbb{R}$ be continuous functions on $\mathbb{T}$. Is f/g continuous on $\mathbb{T}$? State and prove your conjecture.

Supplemental Exercise 16. Let $f : \mathbb{T} \to \mathbb{R}$ be a polynomial function on $\mathbb{T}$. Prove that f is continuous.

Supplemental Exercise 17. Let $f : \mathbb{T} \to \mathbb{R}$ and $g : \mathbb{T} \to \mathbb{R}$ be polynomial functions on $\mathbb{T}$. Is f/g continuous on $\mathbb{T}$? State and prove your conjecture.

Supplemental Exercise 18. Project. Let $f : \mathbb{T} \to \mathbb{R}$ be a function on a time scale $\mathbb{T}$. Prove that f is continuous as in Definition 28 is equivalent to the inverse image, $f^{-1}(\mathbb{S})$, of every time scale $\mathbb{S}$ is either a time scale or the empty set (i.e., it is a closed subset of $\mathbb{T}$).

16

The Delta Derivative

16.1 Delta Differentiation

A study of the behavior of a function $f : \mathbb{T} \to \mathbb{R}$ may begin naturally with an investigation of the behavior of the dependent variable as the independent variable changes. One might ask, can we find a function that describes the relationship between the independent variable and the change in the dependent variable? We call such a function, when it exists, the *delta derivative* of f. So, for a function $f : \mathbb{T} \to \mathbb{R}$, the delta derivative describes the instantaneous rate of change of $f(t)$ relative to the change in $t \in \mathbb{T}$, where the rate exists.

Example 1. Let $f(t) = 3t$ on the time scale $\mathbb{T} = \{2, 4, 6\}$. What is the rate of change of f as t changes from 2 to 4? This rate has a visual interpretation you will probably recognize as the slope of the line segment that joins $(2, f(2))$ and $(4, f(4))$, or

$$\frac{f(4) - f(2)}{4 - 2} = \frac{3(4) - 3(2)}{2} = 3.$$

Using time scale notation, with $t = 2$, this is equivalent to

$$\frac{f(\sigma(t)) - f(t)}{\sigma(t) - t}.$$

What is the rate of change of f for $t = 4$? In this case we get the same rate as when $t = 2$, 3. What about $t = 6$? We cannot calculate the slope there because $t = 6$ is the maximum point in our time scale. We will address this issue in the discussion that follows.

Example 2. Suppose we change the time scale to $\mathbb{T} = [2, 6]$ in the previous example. Calculating

$$\frac{f(\sigma(t)) - f(t)}{\sigma(t) - t}$$

is not possible because because we know that all elements of $\mathbb{T}$ are right dense and therefore, for all $t \in \mathbb{T}$, $\sigma(t) = t$. We cannot calculate a slope at a single point. What we can do is find an expression for slopes of lines through points on the curve near $(t, f(t))$, call them $(s, f(s))$, and then take the limit as $s \to t$. That is, calculate

$$\frac{f(\sigma(t)) - f(s)}{\sigma(t) - s}$$

(the slopes of the secant lines to the curve through the point $(t, f(t))$) and take the limit as $s \to t$. For this example, for $t, s \in \mathbb{T}$, the secant lines have slopes

$$\frac{f(\sigma(t)) - f(s)}{\sigma(t) - s} = \frac{f(t) - f(s)}{t - s}$$
$$= \frac{3t - 3s}{t - s}$$
$$= 3.$$

Now, when we take the limit as $s \to t$, we have

$$\lim_{s \to t} 3 = 3.$$

We did not choose a specific t for our calculations. The rate of change will be 3 for all $t \in \mathbb{T}$. Also, although the limit did not have an effect on our final calculation, we needed to begin with the slopes of the secant lines and pass to the limit because our points were right dense.

We are ready to define the *delta derivative* for a continuous function $f : \mathbb{T} \to \mathbb{R}$. The domain set for the delta derivative, denoted by $\mathbb{T}^\kappa$, is the time scale $\mathbb{T}$ with M removed. When it exists, the delta derivative will be well defined on the set $\mathbb{T}^\kappa$ defined as follows:

Definition 3. Given a time scale $\mathbb{T}$, the set $\mathbb{T}^\kappa$ is

$$\mathbb{T}^\kappa = \begin{cases} \mathbb{T} - \{M\}, & \text{if } \mathbb{T} \text{ has a left-scattered maximum} \\ \mathbb{T}, & \text{otherwise} \end{cases}.$$

Definition 4. Let $f : \mathbb{T} \to \mathbb{R}$ be a function and $t \in \mathbb{T}^\kappa$. The *delta derivative* of f at t, denoted by $f^\Delta(t)$ is defined as

$$f^\Delta(t) = \lim_{s \to t} \frac{f(\sigma(t)) - f(s)}{\sigma(t) - s},$$

if the limit exists.

Let us examine the relationship between continuity and delta differentiability. We consider a function that is delta differentiable at a right-dense point.

Proposition 5. *Let $f : \mathbb{T} \to \mathbb{R}$ be a function where $\mathbb{T}$ is a time scale and $t \in \mathbb{T}$ is a right-dense point. If f is delta differentiable at t, then f is continuous at t.*

Proof. Since f is delta differentiable at t,

$$\lim_{s \to t}(f(\sigma(t)) - f(s)) = \lim_{s \to t}\left[\frac{f(\sigma(t)) - f(s)}{\sigma(t) - s}(\sigma(t) - s)\right]$$
$$= \left(f^\Delta(t)\right)(\sigma(t) - t).$$

Since t is right dense, $\sigma(t) = t$ and

$$\lim_{s \to t}(f(t) - f(s)) = \left(f^\Delta(t)\right)(t - t) = 0.$$

Hence, $\lim_{s \to t} f(s) = f(t)$ and f is continuous at t. □

16.1. Delta Differentiation

Proposition 5 can be generalized to any t, not necessarily right dense. This will be seen in Theorem 14 whose proof will use a new idea in Definition 9. Before proving this, let's look at an example.

Example 6. Let $f : \mathbb{T} \to \mathbb{R}$ be defined by $f(t) = t^2$ on the time scale

$$\mathbb{T} = \{1, 3, 4, 5, 11\} \cup [7, 10].$$

We will calculate the value of $f^\Delta(t)$ for $t = 5$ and $t = 8$. First we determine our set $\mathbb{T}^\kappa$. Since our set has a left-scattered maximum, 11,

$$\mathbb{T}^\kappa = \{1, 3, 4, 5\} \cup [7, 10].$$

For the right-scattered point (in $\mathbb{T}^\kappa$), $t = 5$, we have from the definition

$$\begin{aligned}
f^\Delta(5) &= \lim_{s \to 5} \frac{f(\sigma(5)) - f(s)}{\sigma(5) - s} \\
&= \frac{f(\sigma(5)) - f(5)}{\sigma(5) - 5} \\
&= \frac{7^2 - 5^2}{7 - 5} \\
&= 12.
\end{aligned}$$

For the right-dense point (in $\mathbb{T}^\kappa$), $t = 8$, we have

$$\begin{aligned}
f^\Delta(8) &= \lim_{s \to 8} \frac{f(\sigma(8)) - f(s)}{\sigma(8) - s} \\
&= \lim_{s \to 8} \frac{f(8) - f(s)}{8 - s} \\
&= \lim_{s \to 8} \frac{8^2 - s^2}{8 - s} \\
&= \lim_{s \to 8} \frac{(8 - s)(8 + s)}{8 - s} \\
&= \lim_{s \to 8} 8 + s \\
&= 16.
\end{aligned}$$

For any $t \in \mathbb{T}^\kappa$, regardless if it is right scattered or right dense, we can find an expression for $f^\Delta(t)$.

$$\begin{aligned}
f^\Delta(t) &= \lim_{s \to t} \frac{f(\sigma(t)) - f(s)}{\sigma(t) - s} \\
&= \lim_{s \to t} \frac{(\sigma(t))^2 - s^2}{\sigma(t) - s} \\
&= \lim_{s \to t} (\sigma(t) + s) \\
&= \sigma(t) + t.
\end{aligned}$$

Remark 7. Since f in Example 6 is delta differentiable at the dense point 8, f is continuous at 8 by Proposition 5. Note that

$$\lim_{s \to 8} f(s) = f(8) = 64.$$

For the isolated point, 5,

$$\lim_{s \to 5} f(s) = f(5) = 25.$$

It is easy to see that f is continuous at 5 also.

> **Exercise 8.** For Example 6, calculate the values of $f^\Delta(4)$, $f^\Delta(7)$, and $f^\Delta(10)$.

Directly from our definition of the delta derivative, we have what is known as the "Simple Useful Formula" for calculating, for a delta differentiable function $f : \mathbb{T} \to \mathbb{R}$, the value of $f^\sigma(t)$, which is used as a shorthand for $f(\sigma(t))$.

Definition 9. The *Simple Useful Formula*, denoted by *SUF*, is defined

$$f^\sigma(t) = f(t) + \mu(t) f^\Delta(t),$$

where $f : \mathbb{T} \to \mathbb{R}$ is delta differentiable at $t \in \mathbb{T}^\kappa$.

As the name implies, the formula is straightforward and useful. Moreover, we will show that it holds true for all $t \in \mathbb{T}^\kappa$ where f is delta differentiable.

Example 10. If $\mathbb{T} = \mathbb{R}$, we know that $\sigma(t) = t$ and, therefore,

$$\mu(t) = \sigma(t) - t = t - t = 0.$$

Let f be delta differentiable at t. So, in this case our SUF is not very interesting, but is still true. We have

$$f^\sigma(t) = f(\sigma(t)) = f(t) = f(t) + 0 = f(t) + \mu(t) f^\Delta(t).$$

Example 10 generalizes to right-dense points.

> **Exercise 11.** For right-dense points, $t \in \mathbb{T}$, show that the SUF follows.

When we have a right-scattered point ($\sigma(t) > t$), the formula gives us interesting information. To find the value of $f^\sigma(t)$, what do we do? We begin at the point $(t, f(t))$ and move with slope $f^\Delta(t)$ for the time interval $\mu(t) = \sigma(t) - t$.

> **Exercise 12.** For right-scattered points, $t \in \mathbb{T}$, show that the SUF follows.

The SUF gives you a way to compute $f^\sigma(t)$ indirectly, using $f(t)$, $\mu(t)$, and $f^\Delta(t)$.

> **Exercise 13.** Let $\mathbb{T} = \{3, 4, 8, 11\}$. Suppose that $f(3) = \frac{1}{2}$, $f^\Delta(3) = 1$, $f^\Delta(4) = -1$, and $f^\Delta(8) = 0$. Compute $f(4)$, $f(8)$, and $f(11)$.

16.2. Higher Order Delta Differentiation

Moving across the time scale, the delta derivative changes with the type of point encountered. At a right-dense point, the delta derivative at t, $f^\Delta(t)$, is the limit of the slope of the secant lines,
$$\frac{f(\sigma(t)) - f(s)}{\sigma(t) - s},$$
as $s \to t$. For a right-scattered point, the delta derivative, $f^\Delta(t)$ is the slope of the line segment between $(t, f(t))$ and $(\sigma(t), f(\sigma(t)))$, which is
$$\frac{f(\sigma(t)) - f(t)}{\sigma(t) - t}.$$
This has a wonderful visual interpretation related to slope.

We can now prove that delta differentiability implies continuity.

Theorem 14. *Let $f : \mathbb{T} \to \mathbb{R}$ be a function where $\mathbb{T}$ is a time scale and $t \in \mathbb{T}^\kappa$. If f is delta differentiable at t, then f is continuous at t.*

Proof. Since f is delta differentiable at t,

$$\lim_{s \to t}(f(t) - f(s)) = \lim_{s \to t}\left[\frac{f(t) - f(\sigma(t)) + f(\sigma(t)) - f(s)}{\sigma(t) - s}(\sigma(t) - s)\right]$$

$$= \lim_{s \to t}\left[f(t) - f(\sigma(t)) + \left[\frac{f(\sigma(t)) - f(s)}{\sigma(t) - s}(\sigma(t) - s)\right]\right]$$

$$= f(t) - f(\sigma(t)) + \lim_{s \to t}\left[\frac{f(\sigma(t)) - f(s)}{\sigma(t) - s}(\sigma(t) - s)\right]$$

$$= f(t) - f^\sigma(t) + f^\Delta(t)\mu(t).$$

The last line is 0 by the SUF. Hence, $\lim_{s \to t} f(s) = f(t)$ and f is continuous at t. □

The proofs of Proposition 5 and Theorem 14 are similar except that, in the penultimate step, we use either the fact that t is right dense or the SUF, respectively. The proof of Theorem 14 also proves Proposition 5.

16.2 Higher Order Delta Differentiation

Higher order delta derivatives exist. For example, the second delta derivative of a function $f : \mathbb{T} \to \mathbb{R}$, denoted by $f^{\Delta\Delta}(t)$, is defined in a natural way by

$$f^{\Delta\Delta}(t) = (f^\Delta(t))^\Delta$$
$$= \left(\lim_{s \to t} \frac{f(\sigma(t)) - f(s)}{\sigma(t) - s}\right)^\Delta.$$

We will have the same issue with a left-scattered maximum point for the domain of the second delta derivative as we had with the domain of the first delta derivative. To obtain the domain for the second delta derivative, we start with the domain for the delta derivative, $\mathbb{T}^\kappa$, and remove a left-scattered maximum, if such a point exists. The notation we use for the domain of the second delta derivative is $\mathbb{T}^{\kappa^2}$.

Let's look at some examples, beginning with the linear function, $f(t) = t$.

Example 15. For $f : \mathbb{T} \to \mathbb{R}$ defined by $f(t) = t$, the second delta derivative takes the following form

$$f^{\Delta\Delta}(t) = (f^\Delta(t))^\Delta$$
$$= \left(\lim_{s \to t} \frac{f(\sigma(t)) - f(s)}{\sigma(t) - s}\right)^\Delta$$
$$= \left(\lim_{s \to t} \frac{\sigma(t) - s}{\sigma(t) - s}\right)^\Delta$$
$$= (\lim_{s \to t} 1)^\Delta$$
$$= (1)^\Delta$$
$$= \lim_{s \to t} \frac{1 - 1}{\sigma(t) - s}$$
$$= 0.$$

Therefore, the rate of change of the rate of change of the identity function is zero, or $((t)^\Delta)^\Delta = 0$. This is true for all time scales $\mathbb{T}$ and for all $t \in \mathbb{T}^{\kappa^2}$.

Example 16. For $f(t) = t^2$, we found in Example 6 that the delta derivative is

$$(t^2)^\Delta = \sigma(t) + t.$$

Then, for the second delta derivative, we have,

$$((t^2)^\Delta)^\Delta = (\sigma(t) + t)^\Delta$$
$$= \lim_{s \to t} \frac{\sigma(\sigma(t)) + \sigma(t) - (\sigma(s) + s)}{\sigma(t) - s}$$
$$= \lim_{s \to t} \frac{\sigma(\sigma(t)) - \sigma(s)}{\sigma(t) - s} + \lim_{s \to t} \frac{\sigma(t) - s}{\sigma(t) - s}$$
$$= \lim_{s \to t} \frac{\sigma(\sigma(t)) - \sigma(s)}{\sigma(t) - s} + 1.$$

From this expression, for general classes of t values (right-scattered points or right-dense points) in a given time scale $\mathbb{T}$, we can find a more explicit expression.

In the following two exercises, use the same function and time scale as in Example 16.

Exercise 17. Find $f^{\Delta\Delta}(t)$ for $t \in \{t : t \in \mathbb{T}^\kappa$ and t is right dense$\}$.

Exercise 18. Find $f^{\Delta\Delta}(t)$ for $t \in \{t : t \in \mathbb{T}^\kappa$ and t is right scattered$\}$.

16.3 Properties of the Delta Derivative

Suppose
$$f(t) = 2t - 3t^2.$$

16.3. Properties of the Delta Derivative

Using the Definition of the delta derivative and properties of limits, we can calculate $f^\Delta(t)$ for $t \in \mathbb{T}^\kappa$

Example 19. Let
$$f(t) = 2t - 3t^2$$
for $t \in \mathbb{T}$. Then

$$f^\Delta(t) = \lim_{s \to t} \frac{f(\sigma(t)) - f(s)}{\sigma(t) - s}$$

$$= \lim_{s \to t} \frac{2\sigma(t) - 3(\sigma(s))^2 - (2s - 3s^2)}{\sigma(t) - s}$$

$$= \lim_{s \to t} \frac{2\sigma(t) - 2s}{\sigma(t) - s} - \lim_{s \to t} \frac{3\sigma(t)^2 - 3s^2}{\sigma(t) - s}$$

$$= \lim_{s \to t} 2 \frac{\sigma(t) - s}{\sigma(t) - s} - \lim_{s \to t} 3 \frac{\sigma(t)^2 - s^2}{\sigma(t) - s}$$

$$= 2 \lim_{s \to t} \frac{\sigma(t) - s}{\sigma(t) - s} - 3 \lim_{s \to t} \frac{\sigma(t)^2 - s^2}{\sigma(t) - s}$$

$$= 2 - 3 \lim_{s \to t} \frac{(\sigma(t) - s)(\sigma(t) + s)}{\sigma(t) - s}$$

$$= 2 - 3(\sigma(t) + t).$$

The final line of Example 19 is the delta derivative of our function with respect to t. What can we say about the delta derivative of our function with respect to $f_1(t) = t$ and $f_2(t) = t^2$? The third line from the end of Example 19 can be written in the form of

$$\lim_{s \to t} \frac{\sigma(t) - s}{\sigma(t) - s} - 3 \lim_{s \to t} \frac{\sigma(t)^2 - s^2}{\sigma(t) - s} = 2 f_1^\Delta(t) - 3 f_2^\Delta(t).$$

This leaves us with the impression that the delta derivative of a sum is the sum of the delta derivatives. To verify this, we must prove that this is the case for all delta differentiable functions. With this in mind, we now apply the delta derivative to a linear combination of functions $\alpha f_1(t) + \beta f_2(t)$ where $\alpha, \beta \in \mathbb{R}$, as well as a product, a quotient, or a composition of two or more delta differentiable functions. We will also consider the multiplicative inverse of a delta differentiable function. As with our comment after Example 19, our interest is in the form the delta derivative of our function takes with respect to the functions from which it is constructed.

Theorem 20. Let $f_1, f_2 : \mathbb{T} \to \mathbb{R}$ be delta differentiable functions for $t \in \mathbb{T}^\kappa$. Then the following hold.
(i) $f(t) = \alpha f_1(t) + \beta f_2(t)$ is delta differentiable with delta derivative

$$f^\Delta(t) = \alpha f_1^\Delta(t) + \beta f_2^\Delta(t).$$

(ii) $f(t) = f_1(t) f_2(t)$ is delta differentiable with delta derivative

$$\left(f_1(t) f_2(t)\right)^\Delta = f_1^\Delta(t) f_2(\sigma(t)) + f_1(t) f_2^\Delta(t).$$

(iii) if $f_1(t) f_1(\sigma(t)) \neq 0$ for $t \in \mathbb{T}$, then $f(t) = \frac{1}{f_1(t)}$ is delta differentiable with delta derivative

$$\left(\frac{1}{f_1(t)}\right)^\Delta = -\frac{f_1^\Delta(t)}{f_1(t) f_1(\sigma(t))}.$$

(iv) if $f_2(t) f_2(\sigma(t)) \neq 0$ for $t \in \mathbb{T}$, then $f(t) = \frac{f_1(t)}{f_2(t)}$ is delta differentiable with delta derivative

$$f^\Delta(t) = \frac{f_1^\Delta(t) f_2(t) - f_1(t) f_2^\Delta(t)}{f_2(t) f_2(\sigma(t))}.$$

Proof. Part (i) is left as an exercise.

For Part (ii), using Definition 4 and applying it to

$$f(t) = f_1(t) f_2(t)$$

we obtain

$$f^\Delta(t) = (f_1(t) f_2(t))^\Delta$$
$$= \lim_{s \to t} \frac{f_1(\sigma(t)) f_2(\sigma(t)) - f_1(s) f_2(s)}{\sigma(t) - s}.$$

Now we choose a term to add and subtract, $f_1(s) f_2(\sigma(t))$, and obtain

$$\lim_{s \to t} \frac{f_1(\sigma(t)) f_2(\sigma(t)) - f_1(s) f_2(s)}{\sigma(t) - s}$$
$$= \lim_{s \to t} \frac{f_1(\sigma(t)) f_2(\sigma(t)) - f_1(s) f_2(s)}{\sigma(t) - s}$$
$$+ \lim_{s \to t} \frac{f_1(s) f_2(\sigma(t)) - f_1(s) f_2(\sigma(t))}{\sigma(t) - s}$$
$$= \lim_{s \to t} \left(f_1(s) \frac{f_2(\sigma(t)) - f_2(s)}{\sigma(t) - s} + f_2(t) \frac{f_1(\sigma(t)) - f_1(s)}{\sigma(t) - s} \right).$$

Then, when $s \to t$, we obtain the desired result,

$$\lim_{s \to t} \frac{f_1(\sigma(t)) f_2(\sigma(t)) - f_1(s) f_2(s)}{\sigma(t) - s} = f_1(t) f_2^\Delta(t) + f_2(\sigma(t)) f_1^\Delta(t).$$

For Part (iii), we begin with the definition of the delta derivative of the function

$$f(t) = \frac{1}{f_1(t)},$$

or

$$f^\Delta(t) = \left(\frac{1}{f_1(t)}\right)^\Delta = \lim_{s \to t} \frac{\frac{1}{f(\sigma(t))} - \frac{1}{f(s)}}{\sigma(t) - s}.$$

Simplifying our fraction, we obtain

$$\lim_{s \to t} \frac{\frac{1}{f(\sigma(t))} - \frac{1}{f(s)}}{\sigma(t) - s} = \lim_{s \to t} -\frac{f(\sigma(t)) - f(s)}{f(\sigma(t)) f(s)(\sigma(t) - s)}$$
$$= -\lim_{s \to t} \frac{1}{f(\sigma(t)) f(s)} \frac{f(\sigma(t)) - f(s)}{\sigma(t) - s}$$

16.3. Properties of the Delta Derivative

When $s \to t$, we have

$$\lim_{s \to t} \frac{\frac{1}{f(\sigma(t))} - \frac{1}{f(s)}}{\sigma(t) - s} = -\lim_{s \to t} \frac{1}{f(\sigma(t))f(s)} \frac{f(\sigma(t)) - f(s)}{\sigma(t) - s}$$

$$= -\frac{1}{f(\sigma(t))f(t)} f^\Delta(t),$$

our result.

Part (iv) is left as an exercise. □

Exercise 21. Prove Part (i) of Theorem 20.

Exercise 22. Prove Part (iv) of Theorem 20.

Exercise 23. Prove that the delta derivative of $f_1(t) f_2(t)$ is equivalent to $f_1^\Delta(t) f_2(t) + f_1(\sigma(t)) f_2^\Delta(t)$.

Exercise 24. Using Definition 4, prove that the the second delta derivative of $f(t) = 3t - 2t^2$, on an arbitrary time scale $\mathbb{T}$, is defined by

$$f^{\Delta\Delta}(t) = -2 - 2 \lim_{s \to t} \frac{\sigma^2(t) - \sigma(s)}{\sigma(t) - s}.$$

16.4 Supplemental Exercises

Definition Review. There were three definitions in this chapter. Define each of the following and give an example of each:

Definition 3. for a time scale $\mathbb{T}$, the set $\mathbb{T}^\kappa$

Definition 4. the *delta derivative* of f at t

Definition 9. the *Simple Useful Formula*

Supplemental Exercise 1. Find the delta derivative of $f : \mathbb{T} \to \mathbb{R}$, $f(t) = \sigma(t)$ with

$$\mathbb{T} = \left\{ 1 - \frac{1}{n} \;\middle|\; n \in \mathbb{N} \right\} \cup \{1\}.$$

Write your answer in terms of t.

Supplemental Exercise 2. Find a function that is continuous but not delta differentiable.

Supplemental Exercise 3. Prove that a function is delta differentiable at every isolated point.

Supplemental Exercise 4. Prove that $(fgh)^\Delta = f^\Delta gh + g^\Delta h f^\sigma + h^\Delta g^\sigma f^\sigma$.

Supplemental Exercise 5. Find an expression (different from Supplemental Exercise 4) for $(fgh)^\Delta$.

Supplemental Exercise 6. Find an expression for the delta derivative of $f(t) = at^2 + bt + c$, where $a, b, c \in \mathbb{R}$.

Supplemental Exercise 7. Let $\mathbb{T} = \left\{ 1 + \frac{n}{n+1} \;\middle|\; n \in \mathbb{N} \right\} \cup \{2\}$. Suppose $f(1) = 2$ and $f^\Delta(1) = \frac{2}{3}$. Find an expression for $f(t)$ for $t \neq 2$.

Supplemental Exercise 8. Explore the differentiation of composite functions. For functions on open intervals, the formula for the composite of differentiable functions is known as the Chain Rule.

HINTS

17

Hints for (and Comments on) the Exercises

Hints for Chapter 2

Exercise 22. That is, *if p and q is true, then p is true* and *if p is true, then p or q is true*. Of course, $p \land q \Rightarrow q$ and $q \Rightarrow p \lor q$ are also tautologies. (Treat "p and q" as a single term. Hence, we say "*p* and *q* is true," not "*p* and *q* are true.")

Exercise 23. That is, *if p implies q is true, then p is false or q is true*.

Exercise 24. These are known as the reflexive, symmetric, and transitive properties of equivalence, respectively.

Exercise 25. This is why the phrases "if and only if" and "necessary and sufficient" make sense.

Exercise 27. The point of this exercise is to make sure that you understand the value of truth tables! Every row in a truth table represents an entire paragraph in words.

Exercise 29. Be careful! (d) is impossible!

Exercise 32. You can do this in words or with truth tables. This is an important tool for constructing proofs.

Exercise 34. This is Exercise 32 after a change in variables.

Exercise 35. You can disprove with a counterexample or with truth tables.

Exercise 36. This is the same kind of algebra exercise as you have often seen before.

Hints for Chapter 3

Exercise 3. Mimic the first part of the proof of Proposition 2.

Exercise 8. Mimic the proof of Proposition 7.

Exercise 11. In a proof by cases, consider when x is nonnegative and when it is negative.

Exercise 14. Try proof by contraposition. You may assume that, for integers p and q, $q \neq 0$ implies that $p^2 + q^2 \neq 0$; this is proved in Supplemental Exercise 13.

Exercise 15. Since this is the converse of Supplemental Exercise 9, that does not help to prove it. Try proof by contraposition.

Exercise 16. Break the equivalence into two implications: use direct proof for one and contraposition for the other.

Exercise 20. Both integers are even means x is even and y is even. What is the negation of that?

Exercise 26. If m is divisible by p, then m^n is divisible by p is easier to show than its converse

Exercise 27. Mimic the proof of Theorem 22. The phrase "is even" can be replaced by "is divisible by 2;" this becomes "is divisible by 3." Similarly, the condition $m = 2k$ becomes $m = 3k$.

Exercise 28. Suppose there are $n \geq 2$ people none of whom counts themselves as known. Use proof by contradiction: assume that no two people know the same number of participants. Consider the set of numbers of people that each person knows. One person may know no one. Another may know one person. Show that someone must know no one else while someone else must know every one. Explain why this is a contradiction.

Exercise 41. Where is the first step in the induction?

Exercise 42. Use $\sum_{k=1}^{n+1} \frac{1}{k(k+1)} = \left(\sum_{k=1}^{n} \frac{1}{k(k+1)} \right) + \frac{1}{(n+1)((n+1)+1)}$ in a proof by induction.

Exercise 46. Use complete induction. Given n, n is either prime or the product of two integers strictly between 1 and n.

Exercise 47. Consider 0.

Exercise 48. Consider $f(x) = |x|$.

Hints for Chapter 4

Remember Venn diagrams and logic!

Exercise 8. Is it true that if $x \in \emptyset$, then $x \in X$? Statements such as $\emptyset \subset X$ are sometimes called *vacuously true*.

Exercise 9. Remember that $\subset$ does not mean proper subset.

Exercise 10. You could use $(p \Rightarrow q) \wedge (q \Rightarrow r) \Rightarrow (p \Rightarrow r)$ (Supplemental Exercise 9 of Chapter 2), but this is actually straightforward.

Exercise 13. Use $(p \Leftrightarrow q) \Leftrightarrow (p \Rightarrow q) \wedge (q \Rightarrow p)$ (Exercise 25 of Chapter 2).

Exercise 16. Use Exercises 9, 10, and 13.

Exercise 19. Use Exercises 8 and 13; use the tautology $p \wedge F \Leftrightarrow F$, where F is always false (Supplemental Exercise 36 of Chapter 2).

Exercise 20. Use the tautology $p \wedge p \Leftrightarrow p$ (Supplemental Exercise 25 of Chapter 2).

Exercise 24. Use the tautology $p \vee F \Leftrightarrow p$ (Supplemental Exercise 35 of Chapter 2).

Exercise 25. Use the tautology $p \vee p \Leftrightarrow p$ (Supplemental Exercise 24 of Chapter 2).

Exercise 26. Break the compound statement into $A \cap B \subset A$ and $A \subset A \cup B$. Use Exercise 10 for the second part.

Exercise 33. Use $[a, b) = \{x \in \mathbb{R} \mid x \geq a \text{ and } x < b\}$ and the definition of intersection.

Exercise 36. Use the definitions of $[a, b)$, (a, b), and union.

Exercise 40. Apply the definition with $B = \emptyset$ and $X = A$.

Exercise 42. Consider $B = \emptyset$ and $A = B$.

You have been using Venn diagrams, right?

Exercise 43. Try $\emptyset$, $\{1\}$, and $\{1, 2\}$.

Exercise 46. Remember that $\emptyset$ and X are always subsets of X. You must prove that $\emptyset$ is the only subset of $\emptyset$!

Exercise 47. We have seen that $|\mathcal{P}(\emptyset)| = 1 = 2^0$, $|\mathcal{P}(\mathbb{Z}_1)| = 2 = 2^1$, and $|\mathcal{P}(\mathbb{Z}_2)| = 4 = 2^2$. Power, get it?

Exercise 50. $X \subset A$ or $X \subset B$ implies $X \subset A \cup B$. Is the converse true?

Hints for Chapter 5

Exercise 2. They are, respectively, 3-dimensional space, a 2-dimensional lattice of points, two horizontal parallel lines, and two vertical parallel lines (given that the plane has the standard orientation).

Exercise 3. Assume that $A \times B \neq \emptyset$; assume that $A \neq \emptyset$ and $B \neq \emptyset$.

Exercise 4. Assume that $(A \times B) \cap (B \times A) \neq \emptyset$; assume that $A \cap B \neq \emptyset$.

Exercise 12. This is probably the more familiar definition; it is just a matter of the meaning of $f(x)$.

Exercise 16. Graph the half-parabola $y = \sqrt{x}$.

Exercise 17. Use your knowledge of precalculus or calculus.

Exercise 21. If $y \in f(A)$, then there exists $x \in A$ such that $y = f(x)$.

Exercise 22. Suppose $y \in f(A \cup B)$; suppose $y \in f(A) \cup f(B)$.

Exercise 24. Suppose $y \in f(A \cap B)$.

Exercise 26. Use contraposition.

Exercise 28. This is probably the more familiar definition.

Exercise 31. Use Example 30.

Exercise 32. Compare with Exercise 28; uniqueness comes from one-to-one.

Exercise 34. Unravel $g(f(x))$.

Exercise 37. That is, show $((f \circ g) \circ h)(x) = (f \circ (g \circ h))(x)$. Use Exercise 34 to check the domains and codomains.

Exercise 38. Start with $f : X \to Y$, $g : W \to Z$, $Y \subset W$, and $x_1 \neq x_2$ in X.

Exercise 39. The codomain and range of f equals the domain of g. Determine $g(f(X))$.

Exercise 44. What are the domains and codomains of $f \circ g$ and $g \circ f$?

Exercise 46. Use Exercise 32 to construct an inverse. Compare with Supplemental Exercises 22 and 23.

Exercise 51. If $x_1, x_2 \in f^{-1}(y)$, then $f(x_1) = y = f(x_2)$.

Exercise 52 f cannot be one-to-one; use small finite sets for your counterexample.

Exercise 53 f cannot be onto.

Exercise 57. No! State the domains and codomains explicitly.

Exercise 58. You must show that the domains are equal.

Exercise 59. Be sure that the range of f is not a subset of the domain of g.

Hints for Chapter 6

Exercise 5. $<$ on $\mathbb{R}$ is nonreflexive, transitive, and trichotomous. Anything else?

Exercise 6. This is different from Exercise 5; $\subset$ is different from $\subsetneq$ and the structure of $\mathbb{R}$ is different from that of $\mathcal{P}(U)$.

Exercise 7. You may try to consider relations on small finite sets.

Exercise 10. Check reflexive and nonreflexive.

Exercise 11. Compare the symbols $\prec$ and $\preceq$ with $<$ and $\leq$, respectively; recall that our $\subset$ is denoted $\subseteq$ by some authors.

Exercise 12. $x \preceq y$ and $y \preceq x$ imply $x \preceq x$, which implies $x = x$; $x \prec x$ implies $x \neq x$.

Exercise 16. Connected but not trichotomous contradicts nonreflexiveness.

Exercise 21. Assume there are two candidates for greatest element. Use antisymmetry to show that they must be equal.

Exercise 22. Assume that a greatest element is not maximal. Use nonreflexivity of $\prec$.

Exercise 23. Use connectedness.

Exercise 28. Assume there are two least upper bounds. Use antisymmetry to show that they must be equal.

Exercise 35. See Chapter 7, Section 7.2.

Exercise 41. Use $s \approx t$ to show that $[s] \subset [t]$ and $[t] \subset [s]$.

Exercise 42. Show that if $x \in [s] \cap [t]$, then $[s] = [t]$.

Exercise 46. Check the properties.

Exercise 50. Check the properties.

Hints for Chapter 7

Exercise 3. Show that any function $f : \emptyset \to \emptyset$ satisfies the conditions of a bijection vacuously.

Hints for Chapter 8

Exercise 4. Consider I_A.

Exercise 5. Consider the inverse of a one-to-one correspondence.

Exercise 6. Consider the composition of two one-to-one correspondences.

Exercise 9. Use Exercise 5, Exercise 6, and Theorem 7.

Exercise 10. For (a), extend a one-to-one correspondence $\mathbb{Z}_n \to A$ by $n + 1 \mapsto x$. (b) follows from (a).

Exercise 11. Proceed by induction on $|B|$.

Exercise 16. Compare with Theorem 12.

Exercise 20. For a one-to-one correspondence $f : \mathbb{N} \to D$, set $k = f^{-1}(x)$. Map n to $f(n + 1)$ for $n \geq k$.

Exercise 21. Suppose X is an infinite set. By Theorem 19, there is a denumerable subset D of X. Choose $x \in D$. By the proof of Exercise 20, D and $D - \{x\}$ have the same cardinality. Use the identity function on $X - D$ to complete the construction of a bijection from X onto $X - \{x\}$.

Exercise 22. Compare with Exercise 21.

Exercise 24. By transitivity.

Exercise 25. Define a one-to-one correspondence by, for example, alternating positive and negative.

Exercise 27. Use the one-to-one correspondence from the proof of Theorem 26 to define a one-to-one correspondence by alternating positive and negative.

Exercise 30. See Theorem 23.

Exercise 31. Consider a tangent function.

Exercise 32. Assume $\mathbb{R} - \mathbb{Q}$ is countable. Then $\mathbb{Q} \cup (\mathbb{R} - \mathbb{Q})$ is countable.

Exercise 35. Assume it is countable. Use the diagonal technique used in the proof that $(0, 1)$ is uncountable (Theorem 29).

Exercise 36. Use Exercise 35.

Exercise 37. Assume it is countable. A subset S of $\mathbb{N}$, that is, an element of $\mathcal{P}(\mathbb{N})$ can be represented by a sequence of 0s and 1s where the n-th entry, s_n, in the sequence indicates $n \notin S$ if $s_n = 0$ or $n \in S$ if $s_n = 1$. Use the diagonal technique on the list of such sequences.

Exercise 38. Assume that there is a one-to-one correspondence $f : S \to \mathcal{P}(S)$. Let $A = \{s \in S \mid s \notin f(s)\}$. Show that $A \in \mathcal{P}(S)$ but $A \notin f(S)$.

Hints for Chapter 8

Exercise 14. This is similar to Proposition 12.

Exercise 15. This is similar to Proposition 13.

Exercise 17. Add $-x$.

Exercise 18. This is similar to Exercise 17.

Exercise 24. For example, $f(0) + f(x) = f(0 + x) = f(x)$ for all $x \in X$.

Exercise 26. Show that E has no multiplicative identity element. (Compare with the proof of Proposition 29 in Chapter 10.)

Exercise 27. The identity function is an automorphism. Use the facts that the inverses and composites of isomorphisms are isomorphisms.

Exercise 31. Use the Archimedean Principle, Theorem 30, with $x = 1$ and y is an upper bound on $F_\mathbb{N}$.

Exercise 33. Use Theorem 30.

Hints for Chapter 9

Exercise 8. Use proof by contradiction to show that $\mathbb{N}$ is well ordered: assume S is a nonempty subset with no least element and show by induction that 1 and its successors $\operatorname{Succ}^P(1)$ are all in the complement of S.

Exercise 13. Use $\operatorname{Desc}(m) = \{m + 1, m + 2, m + 3, \ldots\}$ from Lemma 12.

Exercise 15. Use induction.

Exercise 16. Use induction.

Exercise 17. Assume there is another additive identity z. Consider $0 + z$.

Exercise 18. For the inductive step, it is easiest to convert to powers of Succ.

Exercise 19. Use Exercise 13.

Exercise 20. Use proof by contradiction in one direction and Definition 11 in the other direction.

Exercise 21. Use proof by contradiction.

Exercise 23. Use induction on p.

Exercise 24. Show $1m = m$ by induction. Remember to show uniqueness!

Exercise 25. Use induction on m. Show $0m = 0$ by induction. For the inductive step, use Exercise 24.

Exercise 26. To show that $(mn)p = m(np)$, proceed by induction on p.

Exercise 27. Use proof by contradiction in one direction.

Exercise 29. Imitate the first half of the proof for $n < p$.

Exercise 30. This is similar to Theorem 28.

Hints for Chapter 10

Exercise 3. Use (**AC**) and (**AA**) for $\mathbb{Z}_+$ to show that $\sim$ is transitive on $\mathbb{Z}_+ \times \mathbb{Z}_+$.

Exercise 8. For transitivity, suppose that $[m, n] < [p, q] < [r, s]$, show that

$$(m + s) + (p + q) < (r + n) + (p + q),$$

and use (**IX**) for $\mathbb{Z}_+$. Then show trichotomy and apply Exercise 16.

Exercise 12. Imitate Proposition 6

Exercise 15. Use (**AA**) for $\mathbb{Z}_+$.

Exercise 16. Remember to show both existence and uniqueness. Compare with Proposition 12 of Chapter 8.

Exercise 17. It is easier to start with $[m, n]$ than p or $-p$. Remember to show both existence and uniqueness. Also, compare with Proposition 13 of Chapter 8.

Exercise 18. Compare with Exercise 17 of Chapter 8.

Exercise 20. Let $m = [m_1, m_2]$, $n = [n_1, n_2]$, and $p = [p_1, p_2]$. Expand $m + n = m + p$ and use (**AX**) for $\mathbb{Z}_+$.

Exercise 24. Use (**MC**) and (**MA**) for $\mathbb{Z}_+$.

Exercise 25. Use (**MC**) and (**MA**) for $\mathbb{Z}_+$.

Exercise 27. Compare with Exercise 15 of Chapter 8.

Exercise 28. Use Theorem 9.

Exercise 30. Use Theorem 9 and cases for $m > 0$ and $m < 0$.

Exercise 31. Given $p \in \mathbb{Z}_+$, what element of $\mathbb{Z}$ resembles p?

Hints for Chapter 11

Exercise 2. Use (**MC**) for $\mathbb{Z}$ to show that $\sim$ is reflexive and symmetric on $\mathbb{Z} \times \mathbb{Z}^*$; use (**MC**), (**MA**), and (**MX**) for $\mathbb{Z}$ to show that $\sim$ is transitive on $\mathbb{Z} \times \mathbb{Z}^*$.

Exercise 7. Use (**MC**), (**MA**), and (**IX**) for $\mathbb{Z}$.

Exercise 9. Use (**MC**), (**MA**), and (**MX**) for $\mathbb{Z}$ to show that $<$ is transitive; use the trichotomy of $<$ on $\mathbb{Z}$ to show that $<$ is trichotomous.

Exercise 12. Use (**MC**), (**MA**), (**MX**), and (**DP**) for $\mathbb{Z}$.

Exercise 15. Use (**AA**), (**MA**), and (**DP**) for $\mathbb{Z}$.

Exercise 16. Remember to show both existence and uniqueness. Compare with Proposition 12 of Chapter 8.

Exercise 17. Remember to show both existence and uniqueness. Compare with Proposition 13 of Chapter 8.

Exercise 18. Compare with Exercise 17 of Chapter 8.

Exercise 20. Use (**MC**) and (**MA**) for $\mathbb{Z}$.

Exercise 22. Use (**MC**) for $\mathbb{Z}$.

Exercise 23. Use (**MA**) for $\mathbb{Z}$.

Exercise 25. Use (**MC**) and (**MA**) for $\mathbb{Z}$. Also, compare with Exercise 15 of Chapter 8.

Exercise 26. Compare with Exercise 18 of Chapter 8.

Exercise 27. Use (**MC**) and (**MA**) for $\mathbb{Z}$.

Exercise 29. Use (**PP**) for $\mathbb{Z}$.

Exercise 30. Compare with Supplemental Exercise 25 of Chapter 8.

Hints for Chapter 12

Exercise 3. Use contraposition for (a) and (b). Use contradiction for both directions of the equivalence of (a) and (c).

Exercise 5. Use Exercise 3 to show that $<$ is trichotomous.

Exercise 6. Suppose $\mathscr{S} \subset \mathbb{R}$ is nonempty and bounded above by U; that is, $S \subset U$ for all $S \in \mathscr{S}$. Set $L = \cup \mathscr{S}$. Show that $S \in \mathbb{R}$ and S is the least upper bound of $\mathscr{S}$.

Exercise 8. Show that $X + Y$ is a cut using properties of the rationals.

Exercise 9. Use the commutativity of addition of rationals.

Exercise 10. Use the associativity of addition of rationals.

Exercise 11. Show that $X + 0 \subset X$ and $X \subset X + 0$. Similarly for $0 + X$.

Exercise 14. Remember uniqueness!

Exercise 15. Use the Cancellation Property for Addition (Exercise 17) to show strict inequality.

Exercise 17. Show that XY is a cut for $X > 0$ and $Y > 0$.

Exercise 18. Use the commutativity of multiplication of rationals.

Exercise 19. Use the associativity of multiplication of rationals.

Exercise 20. Show that $1X = X$ by chasing elements.

Exercise 21. Let

$$X^{-1} = \{q \in \mathbb{Q} \mid q \leq 0, \text{ or}$$
$$q > 0 \text{ such that } q^{-1} \notin X, \text{ and}$$
$$q^{-1} \text{ is not the least element of } \mathbb{Q} - X \}.$$

Exercise 22–26. Follow the cases of Definition 16 using the result for positive cuts in Exercises 17–21.

Exercise 27. Map $q \in \mathbb{Q}$ to $\{p \in \mathbb{Q} \mid p < q\}$.

Exercise 30. Consider cases: two positives, etc.

Exercise 31. Use $m1^{-1} = m$.

Exercise 32. Use field properties on $(mn^{-1})(pq^{-1})$.

Exercise 33. Show that $\{q \in \mathbb{Q} \mid q < x\}$ is a cut.

Exercise 34. These are one-to-one correspondences.

Exercise 35. $f_\mathbb{Q}$ preserves order.

Exercise 36. Thinking of cuts, $x < y$ implies $f(x) \leq f(y)$. By the density of rationals, there exist $q, r \in \mathbb{Q}$ such that $x < q < r < y$.

Exercise 37. Use Exercise 36.

Exercise 38. For $q \in F$, let $S = \{q \in \mathbb{Q} \mid f(q) < q\}$ and $x = \sup S$. Show that $f(x) = q$.

Exercise 39. If $x, y \in \mathbb{R}$ and $q \in \mathbb{Q}$ such that $f(x + y) < f(q) < f(x) + f(y)$, then there exist $q_1, q_2 \in \mathbb{Q}$ such that $q_1 + q_2 = q$, $x < q_1$, and $y < q_2$.

Exercise 40. Start with positive real numbers.

Exercise 41. Use the properties of isomorphisms to show that $f(x) = x$ for all $x \in \mathbb{Z}$, then for all $x \in \mathbb{Q}$, and finally for all $x \in \mathbb{R}$.

Exercise 42. Consider $g^{-1} \circ f$ for isomorphisms $f, g : \mathbb{R} \to \boldsymbol{F}$.

Hints for Chapter 13

Exercise 5. Consider cases of two positives and otherwise.

Exercise 6. Use the Triangle Inequality, Exercise 5, with the larger of the two Ns.

Exercise 7. You can do this directly or by showing that $\{-y_n\}$ converges to $-y$.

Exercise 8. Show that

$$\begin{aligned}|x_n y_n - xy| &\leq |x_n y_n - x_n y| + |x_n y - xy| \\ &< |x_n||y_n - y| + |y||x_n - x|.\end{aligned}$$

Exercise 9. Show that $\{y_n^{-1}\}$ converges to y^{-1}.

Exercise 10. Show that $|q_m - q_n| \leq |q_m - q| + |q_n - q|$.

Exercise 13. Consider the tail of the sequence, where the terms are within ε of each other— say for $\varepsilon = 1$, but do not forget the first $N - 1$ terms!

Exercise 16. Mimic the proof for increasing sequences.

Exercise 18. Combine Lemma 17 and Theorem 15.

Exercise 21. Use the Triangle Inequality to verify the transitive property.

Exercise 24. Show that if it works for sequences $x \in [x]$ and $y \in [y]$, then it works for sequences $x' \in [x]$ and $y' \in [y]$.

Exercise 26. Show that $<$ is transitive and trichotomous.

Exercise 28. Show that if it works for sequences $x \in [x]$ and $y \in [y]$, then it works for sequences $x' \in [x]$ and $y' \in [y]$.

Exercise 30. Show that if it works for sequences $x \in [x]$ and $y \in [y]$, then it works for sequences $x' \in [x]$ and $y' \in [y]$.

Exercise 31. Things have been put in place.

Exercise 32. Remember that $q_n \in \mathbb{Q}$ and $x \in \mathbb{R}$ is a cut.

Exercise 33. Recall Definition 20.

Exercise 34. Use the denseness of $\mathbb{Q}$ in $\mathbb{R}$.

Exercise 35. Use Definition 27 and the construction of f.

Exercise 36. Use Definition 29 and the construction of f.

Exercise 37. If $[\{x_n\}] \in \mathbb{R}_C$, then $\{x_n\}$ is Cauchy and converges in $\mathbb{R}$, say to x. Show that $f(x) = [\{x_n\}]$.

Hints for Chapter 14

Exercise 4. This follows from the commutativity of addition of real numbers.

Exercise 5. This follows from the associativity of addition of real numbers.

Exercise 6. Try $0 = (0, 0)$.

Exercise 7. Try $-z = (-a, -b)$.

Exercise 8. This follows from the commutativity of multiplication and addition of real numbers.

Exercise 9. This follows from the properties of multiplication and addition of real numbers.

Exercise 10. Try $1 + (1, 0)$.

Exercise 11. Try $\left(\frac{a}{a^2-b^2}, \frac{-b}{a^2-b^2}\right)$.

Exercise 12. This follows from the properties of multiplication and addition of real numbers.

Exercise 13. Yes! Find it.

Exercise 16. Show transitivity and trichotomy.

Exercise 17. To see this is not an ordered field, compare i with 0 and 0 with -1 and consider the product of positives property of ordered fields.

Exercise 18. Either $i > 0$ or $i < 0$. (Why?) Use Supplemental Exercise 26 of Chapter 8.

Exercise 19. Recall the real part of a complex number.

Exercise 20. Consider the image of i and the preservation of multiplication.

Hints for Chapter 15

Exercise 7. Suppose that $b < L$ and choose $0 < \varepsilon < (L - b)/2$. Proceed analogously to Case 1.

Exercise 9. Proceed analogously to the proof of Proposition 8 by showing the existence of a sequence convergence to a.

Exercise 15. Suppose that there is a sequence in the intersection converging to x. To show that x is in the intersection assume that it is not and derive a contradiction.

Exercise 18. Use $\sup \emptyset = \inf \mathbb{T}$ for $\rho(2)$.

Exercise 27. Choose $\delta > 0$ so that $[t_0 - \delta, t_0 + \delta]_\mathbb{T} = \{t_0\}$.

Hints for Chapter 16

Exercise 8. You can use the definition of the delta derivative or use the formula developed in Example 6.

Exercise 11. Imitate Example 10.

Exercise 12. Use Definition 4.

Exercise 13. Use the Simple Useful Formula.

Hints for Chapter 16

Exercise 17. Refer to Example 16.

Exercise 18. Refer to Example 16.

Exercise 21. Use the discussion that proceeded Theorem 20.

Exercise 22. Use Parts (ii) and (iii) of Theorem 20.

Exercise 23. Rework the proof that is given for Part (ii).

Exercise 24. Use Theorem 20 (i) and the delta derivatives of $f_1(t) = t$ and $f_2(t) = t^2$.

Bibliography

[1] David M. Burton, *The History of Mathematics: An Introduction,* 3rd edition, Addison-Wesley, Boston, Mass., 2008.

[2] Michael De Villiers, "The role and function of proof in mathematics," *Pythagoras,* **24** (1990) 17–24.

[3] Apostolos Doxiadis, *Uncle Petros & Goldbach's Conjecture,* Bloomsbury USA, New York, 2000.

[4] Stillman Drake and C. D. O'Malley, "The Assayer" in *The Controversy on the Comets of 1618,* University of Pennsylvania Press, Philadelphia, 1960 (English translation of Galileo Galilei, *Il Saggiatore* (in Italian), Rome, 1623).

[5] Dale M. Johnson, "The problem of the invariance of dimension in the growth of modern topology I," *Arch. Hist. Exact Sci.,* **20** (1979), 97–188.

[6] Victor J. Katz, *A History of Mathematics: An Introduction,* 3rd edition, Addison-Wesley, Boston, Mass., 2008.

[7] Morris Kline, *Mathematical Thought form Ancient to Modern Times,* Oxford University Press, New York, 1972.

[8] Tefcros Michaelides, *Pythagorean Crimes,* Parmenides Press, Las Vegas, 2008.

[9] George Peacock, *A Treatise on Algebra,* J. & J. J. Deighton, Cambridge, England, 1830.

[10] Michael D. Potter, *Set Theory and its Philosophy: A Critical Introduction,* Oxford University Press, New York, 2004.

[11] Bertrand Russell and Alfred North Whitehead, *Principia Mathematica,* Cambridge University Press, Cambridge, England, 1910.

Index

$\mathbb{T}^\kappa$, 200
$\square$, xii
i, 181
Principia Mathematica, 4

absurda, 35, 42, 45, 48, 50
addition
    of complex numbers, 182
    of integers, 148
    of naturals, 139
    of rationals, 157
    of reals, 166, 178
addition-preserving, 128
additive identity element, 124
additive inverse element, 124
algebraic numbers, 5
and, 12
antisymmetric property, 93
arbitrary product of sets, 116
Archimedean Principle, 130
Archimedes of Syracuse (third century B.C.), 129
associative property, 124
    of addition, 124
    of multiplication, 124
assume, 34
asymmetric property, 93
automorphism, 128
axiom, 7
Axiom of Infinity, 38

backward jump operator, 192
bijection, 82
binary operation, 123
Bolzano, Bernard (1781–1848), 176
Bolzano-Weierstrass Theorem, 176
bound
    greatest lower, 98
    least upper, 97
    lower, 98
    upper, 97
bounded sequence, 175

Cantor's real numbers, 177
Cantor's scheme, 113, 115
Cantor, Georg (1845–1918), 3
cardinality
    same, 108
cartesian product, 76
Cauchy sequence, 175
Cauchy, Augustin-Louis (1789–1857), 175
chasing elements, 56
closed interval, 190
closed set, 190
codomain, 78
collection, 56
commutative property, 124
    of addition, 124
    of multiplication, 124
complement of a set, 65
complex numbers, xi, 181
complex numbers,
    addition, 182
    multiplication, 182
composition, 83, 87
    codomain, 87
    domain, 87
    functions, 87
    of functions, 83
    relations, 87
connectedness property, 93
continuous function, 194
contradiction, 17
contrapositive, 20
convergent sequence, 173
converse, 19

countable set, 111

decreasing sequence, 175
Dedekind cut, 165
Dedekind, Richard (1831–1916), 3
delta derivative, 200
dense point, 193
denumerable set, 111
    $\mathbb{Q}$, 113
derivative
    delta, 200
Descartes, René (1596–1650), 76
descendancy set, 139
disjoint sets, 59, 62
    pairwise, 62
distributive properties, 124
domain, 78
Doxiadis, Apostolos (1953–), xiii

element, 56
    chasing, 56
elements
    chasing, 56
embedding, 129
empty set, 56
equal
    sets, 57
equivalence class, 99
equivalence relation, 98
Erdős, Paul (1913–1996), 33
Euclid of Alexandria (fourth century B.C.), 6
Euclid of Alexandria (fourth century B.C.), xii
Eudoxus of Cnidus (fourth century B.C.), 3

Fibonacci
    sequence, 43
Fibonacci, Leonardo (c. 1170–c. 1250), 4
field, 125
    ordered, 125
finite set, 109
for all, 23
forward jump operator, 192
function, xi, 78, 189
    bijective, 82
    continuous, 194
    identity, 85
    injective, 82
    invertible, 85
    surjective, 82
Fundamental Theorem of Arithmetic, 36

Galileo Galilei (1564–1642), 75
greatest element, 96
greatest lower bound, 98

Hilger, Stefan (1959–), 189

identity element
    of addition, 124
    of multiplication, 124
identity function, 85
if —, then —, 14
if and only if, 16
iff, 16
image, 79
imaginary numbers, 183
immediate predecessor, 139
immediate successor, 138
implication, 14
increasing sequence, 175
infimum, 98
infinite product of sets, 116
infinite set, 110
injection, 82
integers, xi, 146
    mod n, xi
    negative, 147
    nonnegative, xi
    nonzero, xi
    positive, 147
integers as differences, 146
integers,
    multiplication, 149
    addition, 148
    ordering, 146
intersection, 58, 61
    of a collection of sets, 61
    of two sets, 58
interval, xi
    closed, xi, 190

Index

half-closed, xi
half-open, xi
open, xi
inverse (of implication), 20
inverse element
    of addition, 124
    of multiplication, 124
inverse function, 85
inverse image, 86
invertible function, 85
irrational number
    $\sqrt{2}$, 35
irrational numbers, 4
is an element of, 56
isolated point, 193
isomorphism, 128

jump operator, 192
    backward, 192
    forward, 192

least element, 97
least upper bound, 97
least upper bound property, 98
left-dense point, 193
left-scattered point, 193
lexicographic ordering, 183
limit, 190
limit of a function, 193
logical statement, 9
lower bound, 98

Main Theorem, 130, 169
mathematical statement, 9
maximal element, 96
minimal element, 97
mod by a partition, 101
mod by an equivalence relation, 100
multiplication
    of complex numbers, 182
    of integers, 149
    of naturals, 141
    of rationals, 158
    of reals, 168, 178
multiplication-preserving, 128
multiplicative identity element, 124

multiplicative inverse element, 124
multiset, 56

natural numbers, xi, 138
naturals, 138
naturals,
    multiplication, 141
    addition, 139
    ordering, 139
necessary and sufficient, 16
negation of quantifiers, 24
nonreflexive property, 93
not, 11

one-to-one, 82
onto, 82
or, 13
order of operations, 123
order-preserving, 128
ordered field, 125
ordered pair, 76
ordering
    lexicographic, 183
    of the integers, 146
    of the naturals, 139
    of the rationals, 156
    of the reals, 166, 177

pairwise disjoint sets, 62
paradox
    Russell's, 66
partial ordering, 95
partially ordered set, 95
partition, 100
Peano Axioms, 137
Peano, Giuseppe (1858–1932), 4, 137
point
    dense, 193
    isolated, 193
    left dense, 193
    left scattered, 193
    right dense, 193
    right scattered, 193
poof, 35
poset, 95
power set, 65

predecessor
    immediate, 139
predicate, 22
preserving,
    addition, 128
    multiplication, 128
    order, 128
prime number, 36
primitive term, 7
Principle
    Archimedean, 130
Principle of Complete Induction, 42
Principle of Induction, 38
product of sets
    arbitray, 116
proof, 7
proof techniques, 30
    direct proof, 30
    proof by brute force, 30
    proof by cases, 33
    proof by complete induction, 43
    proof by contradiction, 34
    proof by contraposition, 33
    proof by counterexample, 46
    proof by example, 46
    proof by induction, 39
Pythagoras of Samos (sixth century B.C.), 3

Q.E.D., xii
quantifier
    existential, 24
    universal, 23

range, 80
rational numbers, xi, 156
    nonnegative, xi
    positive, xi
rationals, 156
    negative, 157
    positive, 157
rationals as quotients, 156
rationals,
    multiplication, 158
    addition, 157
    ordering, 156

real number, 165
real numbers, xi, 177
    nonnegative, xi
    positive, xi
reals,
    multiplication, 168, 178
    addition, 166, 178
    ordering, 166, 177
reflexive property, 93
relation, 77
    antisymmetric, 93
    asymmetric, 93
    connected, 93
    equivalence, 98
    nonreflexive, 93
    on a set, 77, 93
    reflexive, 93
    symmetric, 93
    transitive, 93
    trichotomous, 93
restriction, xi, 81
right-dense point, 193
right-scattered point, 193
Russell's paradox, 66
Russell, Bertrand (1872–1970), 4, 106

sequence, 173, 189
    bounded, 175
    Cauchy, 175
    convergent, 173
    decreasing, 175
    increasing, 175
sequence in a set, 189
set, 56
    closed, 190
    countable, 111
    denumerable, 111
    difference, 64
    empty, 56
    finite, 109
    infinite, 110
    largest, 64
    uncountable, 114
    universal, 64
set equality, 57

sets
    disjoint, 59, 62
    pairwise disjoint, 62
statement, 9
    biconditional, 16
    conditional, 14
    conjunction, 12
    disjunction, 13
    equivalence, 16
    negation, 11
strict partial ordering, 95
strict total ordering, 96
strict totally ordered set, 96
strictly partially ordered set, 95
subset, xi, 56
    proper, xi, 56, 58
successor
    immediate, 138
suppose, 34
supremum, 97
surjection, 82
symmetric property, 93

tautology, 17
Thales of Miletus (sixth century B.C.), 6
there exists, 24
time scale, 189, 191
    interval, 193
total ordering, 96
totally ordered set, 96
transcendental number, 119
transitive property, 93
trichotomy property, 93
true
    vacuously, 15, 30

uncountable set, 114
undefined term, 7
union, 59, 63
    of a collection of sets, 63
    of two sets, 59
unity, 124
universal set, 64
upper bound, 97

vacuously true, 15, 30, 212

Weierstrass, Karl (1815-1897), 176
well-ordered set, 98
well-ordering, 98
Whitehead, Alfred North (1861-1947), 4

zero, 124, 138

About the Authors

Ralph W. Oberste-Vorth was born in Brooklyn, New York and attended New York City public schools. His interests in mathematics began in junior high school and developed at Stuyvesant High School and Hunter College of the City University of New York. He met Aristides Mouzakitis at Hunter, where they became good friends while earning BA and MA degrees in mathematics. Ralph continued his studies at Cornell University where he earned his PhD in dynamical systems under the direction of John Hamal Hubbard. After one-year positions at Yale University and the Institute for Advanced Study in Princeton, New Jersey, he moved to the University of South Florida in 1989. In 2000, Ralph approached Aristides with the idea of writing a "proofs text." During an intense 19-day session at his home in Corfu, Greece, they wrote the first draft of this book. Several instructors at South Florida used it. Ralph moved to Marshall University as the Chairman of the Department of Mathematics in 2002. In 2009, he invited his Marshall colleague, Bonita Lawrence, to help them put the book into a publishable form. The book had been used several times at Marshall. This project was started during the summer of 2009 and completed in the summer of 2011, with input from the MAA. In August 2011, Ralph became the Chairman of the Department of Mathematics and Computer Science at Indiana State University. Ralph lives in Terre Haute, Indiana with his wife and three children.

Aristides Mouzakitis was born in the village of Avliotes on the Greek island Corfu in the Ionian Sea. He attended primary school in Avliotes and moved to Kerkyra, the main town of Corfu, to attend high school. In 1980, Aristides moved to New York City to attend Hunter College of the City University of New York. There, he earned his BA in the Special Honors Curriculum and his MA in mathematics. Aristides met Ralph Oberste-Vorth while at Hunter, where they laid the foundations for an enduring friendship. He began doctoral studies in mathematics, but Greece beckoned. In Greece, he has worked as a teacher in secondary education and as an English - Greek translator of popular mathematics books and articles. Eventually, he took on further formal studies and in 2009 he earned his doctorate in mathematics education from the University of Exeter in England under the direction of Paul Ernest. Aristides stays active in the Astronomical Society of Corfu and the Corfu branch of the Hellenic Mathematical Society. He lives in Kerkyra with his wife and his daughter, and enjoys reading and swimming, especially in winter time.

Bonita Lawrence is currently a Professor of Mathematics at Marshall University in Huntington, West Virginia. She was born to a military family when her father was stationed with the U.S. Army in Stuttgart, Germany. Her father retired at Ft. Sill near Lawton, Oklahoma when she was in junior high school. She received her baccalaureate degree in Mathematics

Education from Cameron University in Lawton in 1979. After ten years of teaching, she returned to school to study for a Master's degree in Mathematics at Auburn University. Upon completion of her Master's degree in 1990, she continued her academic training at the University of Texas at Arlington, earning a Ph.D. in Mathematics in 1994.

In her first teaching position after completing her Ph.D., at North Carolina Wesleyan College, she was the 1996 Professor of the Year. After a few years at small institutions, North Carolina Wesleyan College and the Beaufort Campus of the University of South Carolina, she made the move to Marshall University to expand her teaching opportunities and to work with graduate students at the Master's level. She served as either Associate Chairman or Assistant Chairman for Graduate Studies for 10 of the 11 years under the leadership of Dr. Ralph Oberste-Vorth.

During her time at Marshall University, she has received the following research and teaching awards: Marshall University Distinguished Artists and Scholars Award—Junior Recipient for Excellence in All Fields (Spring 2002); Shirley and Marshall Reynolds Outstanding Teaching Award (Spring 2005); Marshall University Distinguished Artists and Scholars Award—Team Award for Distinguished Scholarly Activity, with one of my coauthors, Dr. Ralph Oberste-Vorth (Spring 2007); Charles E. Hedrick Outstanding Faculty Award(April 2009); and the West Virginia Professor of the Year (March 2010).

Dr. Lawrence currently is the lead researcher for the Marshall Differential Analyzer Lab, a mathematics lab that houses the Marshall Differential Analyzers. These machines, built by students of replicated Meccano components, are models of the machines that were first built in the late 1920's to solve differential equations. The largest of the machines, a four integrator model that can run up to fourth order equations, is the only publicly accessible machine of its size and type in the country. The lab offers the opportunity for the investigation of new research ideas as well as educational experiences for students of mathematics at many levels.

This is her first book as a coauthor. She served as a reviewer for a linear algebra textbook and the solutions manual, *The Keys to Linear Algebra*, by Daniel Solow.

Dr. Lawrence shares her life with her husband of 15 years, Dr. Clayton Brooks, a colleague in the Mathematics Department.